Intergovernmental Oceanographic Commission technical series

22



## The IOC/WMO/UNEP Pilot Project on Monitoring Background Levels of Selected Pollutants in Open Ocean Waters

# Scientific report of the intercalibration exercice

Unesco

*The present report was prepared by N. Andersen, J.M. Bewers, J. Duinker, A. Knap, K. Palmork and J-P. Villeneuve, as an evaluation of the scientific data obtained during a Workshop on the Intercalibration of Sampling Procedures of the IOC/WMO/UNEP Pilot Project on Monitoring Background Levels of Selected Pollutants in Open-Ocean Waters held in Bermuda, 11-26 January 1980, and in the course of specific follow-up activity. The Intergovernmental Oceanographic Commission wishes to acknowledge gratefully the support received from its co-sponsors of the above-mentioned Workshop, the UN Environment Programme (Project FP/0501-79-02) and the World Meteorological Organization, as well as the Bermuda Biological Station for Research, which provided the laboratory facilities and accommodation for the participants in the Workshop, the International Atomic Energy Agency's Internationl Laboratory of Marine Radioactivity, for its specialized technical assistance in the intercalibration of organochlorine standard solutions, the US National Oceanic and Atmospheric Administration, for the provision of the R.V. GEORGE B. KELEZ and some specialized equipment, and the Netherlands Institute for Sea Research, Den Burq, Texel, the Institute of Marine Research, Bergen Norway and the Bermuda Biological Station for Research, Inc., where follow-on experimentation was conducted.*

ISBN 92-3-102077-3

Published in 1982
by the United Nations Educational,
Scientific and Cultural Organization,
7, place de Fontenoy, 75700 Paris

Composed by SEP 2000, Paris
Printed in Unesco's workshop

# Contents

# Introduction and Executive Summary

**N.R. Andersen,**
*Scripps Institute of Oceanography University of California, San Diego, La Jolla, California*
**J.M. Bewers,**
*Bedford Institute of Oceanography, Dartmouth, Nova Scotia*
**K.H. Palmork,**
*Institute of Marine Research, Bergen, Norway*

## Background

The interest of the Intergovernmental Oceanographic Commission (IOC) in monitoring open ocean pollutants can be traced back about a decade. An IOC Intergovernmental Coordination Group (ICG) for the Global Investigation Pollution Marine Environment (GIPME) was established in 1972 and charged with developing a Comprehensive Plan for GIPME ; this task was completed in 1976, and the ICG was transformed into the IOC Working Committee for GIPME. In the development of programmes under this Comprehensive Plan, the IOC through Unesco, requested the IMCO/FAO/Unesco/WMO/WHO/IAEA/UN Joint Group of Experts on the Scientific Aspects of Marine Pollution (GESAMP) to define the Scientific Bases for the Determination of Concentrations and Effects of Marine Pollution (GESAMP Reports and Studies N° 1). The GESAMP Working Group set up for this purpose, proposed an open-ocean marine pollution monitoring system.

In the development of the United Nations Environment Programme's (UNEP) Global Environmental Monitoring System (GEMS), the IOC proposed the implementation of the GESAMP proposal, with the collaboration of WMO, IOC and UNEP developed the IOC/WMO/UNEP Pilot Project for Monitoring Background Levels of Selected Pollutants in Open Ocean Waters. The Executive Committee of WMO approved the project in principle, as amended by its Resolution 16 (EC-XXVIII). The Executive Council of IOC, by its Resolution EC-VII.11, approved the general concept and intention of the proposed project and by Resolution EC-VIII.9 approved the joint IOC/WMO/UNEP Pilot Project for Monitoring Background Levels of Selected Pollutants in Open Ocean Waters.

The long-term objective of this project is to obtain information on the long-term changes and trends in the background levels, in the ocean, of the more common pollutants that can endanger human health, either directly of through a harmful effect on living organisms, or that influence the exchange of energy and matter between the ocean and the atmosphere. The pollutants were selected on the following criteria :

i) that they were known or suspected to have the atmosphere as their major pathway to the open ocean ;
ii) they were introduced by other direct routes such as dumping or deliberate operational discharge from ships ; or
iii) they were sufficiently persistent compounds which, although entering the marine environment mainly via discharge from land, were liable to reach the open ocean by current transport.
iv) are harmful for living resources of the sea.

On the basis of these criteria, the following categories of pollutants were identified (in no particular order of priority) :

1. Heavy Metals
2. Organochlorines
3. Petroleum Hydrocarbons
4. Surface-Active Compounds
5. Radionuclides
6. Solid waste

As a preparatory stage in this proposed international venture, the Pilot Project consists of the following parts :

i) A Preliminary Phase to develop techniques and expertise, and to survey conditions for the selection of suitable monitoring sites, this phase to culminate in a first Workshop, and
ii) A Monitoring Phase to carry out Pilot Monitoring of the Atlantic Ocean, culminating in a second Workshop.

These activities are aimed at assessing and improving the sampling, analytical and interpretative procedures that may be applied within the GIPME programme proper and related programmes. The results of the intercalibration experiment that are reported in this document can be considered as contributing to the Preliminary Phase.

By Resolution EC-VIII.3 on the Recommendation 7 of the First Session of the Working Committee for GIPME (18-22 October 1976), the IOC Executive Council at its Eighth Session (4-8 April 1977) decided to establish the GIPME Group on Experts of Methods, Standards and Intercalibration (GEMSI). At the first session of GEMSI, convened in Paris, 5-9 September 1977, deliberations began on preparing for an open-ocean intercalibration experiment for selected heavy metals and organochlorines (i.e., PCBs) in seawater samples. Development of guidelines for the experiment continued at the Second Session of GEMSI (Bergen, 1-4 May 1978), with specifics of the exercice being finalized at two GEMSI Core Group meetings held at Monaco 12-14 March 1979 and Bermuda 8-10 October 1979. It was concluded that the Bermuda Biological Station for Research, Inc. and Panulirus Station (32°1 ON ; 64°3 OW) would be ideally suited for the land-based laboratory and open-ocean location, respectively (See Fig. 0.1). The scientific basis of these conclusions is presented in IOC Workshop Report No. 25

This report describes the exercise which took place in Bermuda from 11 to 26 January 1980. Thirty-five scientists from thirteen countries participated in the workshop. Most of the participants appear in the photograph in Figure 0.2. The U.S. National Oceanic and Atmospheric Administration provided the R.V. GEORGE B. KELEZ for the open-ocean sampling operations (See Fig. 0.3).

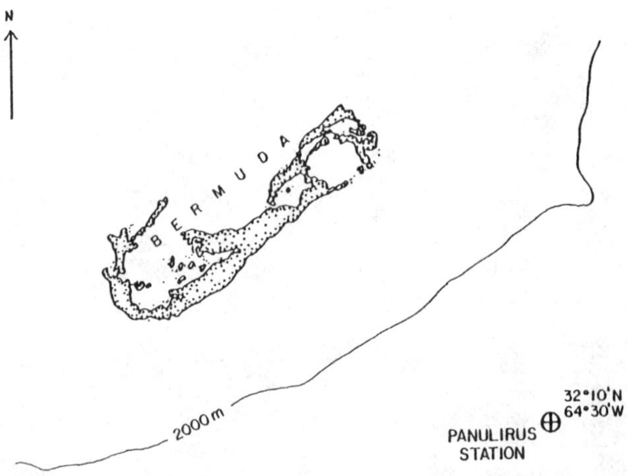


**Figure 01 Panulirus Station Location**

The exercise was divided into two separate operational activities, one concerned with heavy metals, the other with organochlorines. The final results of these two components of the intercalibration exercise are individually pre-

**Figure 02 Participants of the workshop**

sented in the present report, which was approved by the Third Session of GEMSI, Monterrey, California, 8-12 September 1980.

The sea-surface microlayer, as the boundary between the oceans and atmosphere, was considered by the GEMSI Core Group meeting in Bermuda with a view to use in the monotoring of open ocean waters because of :

i) the rapidity with which it can collect and concentrate pollutants ;

ii) the enrichment of the microlayer by pollutant compounds relative to the underlying open-ocean water.

As a result, a demonstration of screen sampling was presented by Dr. Robert Pellenbarg (Naval Research Laboratory, Washington DC) at Bermuda. However, it was felt that the ultimate type of sampler used should be standardized before implementation of an intercalibration exercise and, therefore, the surface film was not included as an operational component of the Bermuda exercise. This topic is presently the subject of intersessional consideration by GEMSI.

Additionally, during the period 13-19 January 1980, two US satellites were scheduled to measure environmental conditions around Bermuda. TIROS-N was to measure water colour (i.e. essentially a measure of the Coastal Zone Colour Scanner (CZCS), with results expected to provide insight into any unusual levels of productivity present during the sampling period. The Advanced Very High Resolution Radiometer (AVHRR) on NOAA-6 was to record surface temperature. These results were looked to for information on any unusual circulation patterns (e.g., eddies) that occurred. Sea-truth measurements were made to complement the remotely sensed data. Unfortunately, the cloud cover over the sampling location was intense and long-term so that, in spite of considerable data manipulation using the Remote Sensing Facility of the Scripps Institution of Oceanography, useful images could not be obtained. Nevertheless, for any future intercalibration endeavours, remotely sensed measurements of the type scheduled in the Bermuda exercise should be included.

## Summary and recommendations trace metals component of the intercalibration exercise

In general, the data returned by the participants in the Trace Metal component of the exercise are of good quality. Indeed, the relative agreement between experienced laboratories is better than previous seawater trace-metal intercomparisons with the exception of the most recent intercalibrations for cadmium and mercury conducted by ICES on behalf of the Joint Monitoring Group of the Oslo and Paris Commissions. The Bermuda exercise was, however, the first international intercalibration to assess directly the influences of commonly-used sampling devices and hydrowires on the determinations of concentrations of several trace metals in open-ocean waters. The largest bodies of data accumulated during the exercise are for cadmium, copper, nickel and zinc, with only a relatively few participants having returned data for iron, lead, manganese, mercury, molybdenum and vanadium. With the exclusion of outlying values, of which there were generally only a few for each element, the overall mean concentrations for all metals, except lead, are comparable with contemporary consensus values for metal concentrations in intermediate and deep open-ocean waters. This is particularly remarkable since several of the participants had only limited experience in the analysis of some of the metals in offshore seawater, for which they have reported data in this experiment.

Large differences in reported values for trace-metals of open-ocean waters have, in the past, often been attributed to the effects of sampling devices and their methods of deployment. The results of this experiment reveal, however, that differences resulting from the application of differing sampling procedures, assuming that reasonable precautions are taken, are relatively small in the cases of most of the commonly analysed trace-metals. With a few exceptions, alternative sampling bottles and hydrowires in common use by experienced marine laboratories do not give rise to large discrepancies in the results of metal analysis of samples collected with them. For some elements there do appear to be significant differences in the results obtained with different sampling procedures. For copper and nickel, HYDRO-BIOS bottles appear to yield greater concentrations than do modified GO-FLO samplers. Similarly, mo-difications made to ex-works GO-FLO samplers, to improve integrity, do result in marked reductions in the concentrations of copper, nickel and zinc measured in seawater samples. For these same metals, and perhaps also for iron, plastic-coated steel would seem to be the cleanest hydrowire of the three types compared. Except in the case of zinc, NISKIN bottles that have been suitably modified appear to be capable of collecting samples of only slightly interior integrity to those collected with modified GO-FLO samplers. Some of these conclusions are supported by the results of an ancillary experiment in which changes in trace-metal concentrations in seawater samples retained within individual bottle types were measured.

The results of the entire exercise indicate that sufficient competence in trace-metal determinations in seawater exists among experienced marine laboratories to justify making available their experience to other laboratories wishing to develop comparable expertise. The experiment has also shown that sampling methods currently being used by most developed marine laboratories are adequate for trace-metal sampling in the ocean at concentrations now believed to prevail in open-ocean waters. It is therefore recommended that IOC, with the collaboration of UNEP and other concerned international organizations, as appropriate, now consider convening other regional workshops in which analysis and trace-metal geochemistry be made available to personnel and institutions in developing coun-

**Figure 03 The R.V. George B. Kelez which participated in the open-ocean operations of the workshop**

tries. This might provide a most expedient mechanism for the provision of such assistance and more rapidly provide widespread capability for the assessment of trace-metal baselines and contamination in regional seas and continental shelf environments.

## Organochlorine component of the intercalibration exercise

This component of the intercalibration exercise was also extremely successful, but for reasons markedly different from those noted above with regard to trace metals.

In this component of the exercise, considerations were focused mainly on chlorinated biphenyls. These compounds pose the largest analytical problems ; other compounds looked for could hardly be traced and identified without the use of gas chromatography-mass spectrometry (GC-MS) techniques. Such capabilities were not available at Bermuda.

Literature on PCBs in seawater, being very limited indeed, has reported concentrations until now almost exclusively in terms of Aroclor 1245 (or the like) equivalent. However, the report on this component of the exercise is one of the very first — and to our knowledge the first by an international body — openly rejecting this method. At the same time, it demonstrates that not only is the method that makes measurements based on individual PCB components (i.e., the one used in Bermuda) the only scientifically meaningful method, but that it is feasible for measuring even the extremely low levels found in solution in open-ocean water.

The success of this component of the intercalibration exercise is attributed to various causes. The presence of the most up-to-date GC instrumentation allowing the application of temperature-programmed glass-capillary gas chromatography has been shown through this work to be essential. The presence of representatives of the various instrument companies ensured « state-of-the-art » performance. It was felt that the availability of a device that records retention times, while obtaining chromatograms, greatly assisted in the interpretation of results.

It is *essential* that capillary-column temperature-programmed gas chromatography be used so as to obtain maximum separation of peaks in the chromatograms, allowing for the precise and accurate determination of individual PCB components, and other well defined organochlorine compounds.

The excellent laboratory facilities and the staff at the Bermuda Biological Station for Research, Inc. of course played a significant role in ensuring success. Also the extensive work performed in laboratories that had volunteered to undertake such work in the period between January and September 1980 significantly enhanced the results obtained in Bermuda. The facilities offered for meetings by the IAEA International Laboratory of Marine Radioactivity (ILMR) in Monaco and the Netherlands Institute for Sea Research in Texel in this period also greatly assisted in the rapid evolution of ideas.

It was decided at Bermuda to use, as far as possible, temperature — programmed glass-capillary-gas chromatography so as to allow quantification of PCBs in terms of individual components. However, at that stage, not all the participating laboratories were adequately equipped for this purpose. Nevertheless, the experience gained at Bermuda has accelerated in the meantime the acquisition of appropriate equipment by several laboratories.

Analysis of the XAD-2 extracts in the home laboratories have been performed using capillary-column GC by practically all participants, in addition to packed-column analyses by three participants.

In spite of the fact that the operation was considered successful on the whole, the comparison of the various sampling techniques was not totally realized. This is attributed to several causes. As pointed out previously, as a consequence of the low levels of PCBs found in the water sampled, a fundamental question arose concerning the validity of the way in which PCBs were quantified using some particular standard formulation (i.e., Aroclor 1245). As a result, a redirection in the major thrust of the operation was made, which shifted the emphasis to analytical considerations. Secondly, the weather conditions presented severe difficulties for the sampling phase of the experiment. It was not possible to work up one set of samples prior to the next sampling day. In retrospect, a longer period for shipboard activity would have been desirable.

The experience and results obtained from the analyses of the XAD-2 extracts in the home laboratories after the Bermuda experiment allow some final conclusions to be made on the feasibility of making reliable estimations of the PCB content in open-ocean waters.

Despite the active co-operation of several specialists in the field, some serious problems were encountered, some of which were solved. However, several problems still need considerable research in the near future, in addition to the work already going on in several countries.

The changes in the sample composition that are most difficult to detect and to eliminate probably occur during sampling and other shipboard activities. Further work is necessary to design appropriate techniques for obtaining and handling sea-water samples at the extremely low concentration levels of PCB and other organochlorines found in open-ocean waters. This will include sampling devices, routine handling of equipment, necessity of clean labs, and eliminating sources of contaminants during normal shipboard activities. It is also important to determine the amounts of any particular compound being in solution and in particulate form. It is as yet not clear to what extent the various techniques used during the exercise are the appropriate means for distinguishing between these forms. A considerable amount of work in various parts of the world's ocean is necessary to unravel these problems ; their solution is essential before any type of monitoring of PCBs in the ocean, either regionally or globally, can be initiated. The requisite research should be encouraged by the appropriate international organizations.

Further work is necessary to identify the individual PCB components present in seawater from different parts of the ocean, with application of GC-MS techniques to extracts of large volumes of seawater. Work of this kind in the Member States should be strongly encouraged.

The results of the exercise at Bermuda did not allow a proper comparison of the efficiencies of XAD-2 and other extraction techniques. However, follow-up experiments during the intersessional period at some laboratories (e.g., Bermuda, Biological Station for Research, Institute of Marine Research in Bergen and Netherlands Institute of Sea Research in Texel) suggest that both resin collection and liquid-liquid extraction techniques are feasible sampling techniques for the analysis of organochlorines in open-ocean if appropriate measures are taken to eliminate interfering peaks. This can be checked by running blanks. The samples collected for solvent extraction were exposed to the atmosphere. However, analysis of the atmospheric content of PCB in the laboratory air at Bermuda during the exercise (performed by the Texas A&M University laboratory) shows that the atmosphere cannot have been a significant source of the interfering peaks.

Based on the experience gained in Bermuda, it is felt that a baseline of the Pilot Project for organochlorines in open ocean waters is feasible, and it should consist of several phases. It is specifically recommended that a two-year programme be initiated. It should start with the collection of samples by different techniques in open-ocean waters. Bermuda has indicated its interest in this pilot phase, and it is highly recommended that the Bermuda Biological Station for Research Inc. be invited to host the activities. This would allow the use of a locally available smaller ship for a longer period of time in periods of more desirable weather conditions (e.g., April-December), than was experienced during this experiment. In the meantime, the Institute of Marine Research (Bergen) has supplied a continuous water extractor to some participants, in order to accelerate research on the comparison of solvent and XAD-2 resin extraction techniques. These attempts should form the basis for an evaluation of the most appropriate methods for sampling and analysis of background levels of organochlorines in open-ocean waters as well as in coastal waters.

The institutes mentioned above have also volunteered to continue the research on extraction in conjunction with the

Bermuda Biological Station for Resarch, Inc. The results obtained after an initial six-month period at Bermuda should be evaluated, possibly including some additional practical work during the two — or three-week meeting of the participants. The results would lead to a formulation of the technique to be applied in the next phase (i.e., an 18-month period of sampling and analysis of organochlorines in open ocean waters of Bermuda). The results obtained over this period should be evaluated in close connection with the data on atmospheric deposition, rainfall etc., that are currently obtained in the acid rainfall programme at Bermuda Biological Station for Research, Inc., and in other programmes.

The operational aspects of the intercalibration exercise have also provided a basis for certain recommendations to be formulated for future activities. Ample time must be allowed for the planning of the exercise, so that participants can be notified as to the specifics of the programme and all equipment can be transported. The participants in such a programme should arrive well before the first sampling day to ensure familiarity with techniques and the host laboratory. Blanks should be run and familiarity with the instrumentation present should be gained.

The participation of highly trained personnel from the instrument companies whose instruments are being used is essential ; the companies provide « state-of-the-art » analytical equipment, and the trained personnel to ensure maximum performance of the instruments.

There should be a feedback mechanism between analytical work-up and ship activities, which enables sampling to occur after preliminary analyses. This would allow modification of experimental design based on hard data.

Co-operation of the work towards the scientific goals and the normal shipboard activities should be done by the Chief Scientist during the cruise as well as on « lay » days in port. General shipboard activities such as painting and lubricating must be controlled in order to prevent contamination of the samples.

It is essential to have all laboratory equipment on site, with workshop and vessel facilities being able to respond to short-term requirements dictated by laboratory results. The laboratory staff should have experience in open-ocean marine pollution chemistry, as was the case in this exercise.

There is no doubt that the degree of success that has been achieved in this experiment has been the result of the continued, dedicated efforts of many individuals in the execution and in the planning. To all these people we are indebted and extend our appreciation. The U.S. National Oceanic and Atmospheric Administration and the Bermuda Biological Station for Research, Inc. deserve special recognition for the ship and shore-based facilities they provided. Particular acknowledgement is due to the international organizations involved, specifically IOC, WMO and UNEP, the support from which made the experiment possible, and for the specialized laboratories of the Netherlands Institute for Sea Research in Texel, the Institute of Marine Research in Bergen and the International Laboratory of Marine Radioactivity (IAEA) in Monaco.

# Chapter I
# Trace Metals

J. Michael Bewers,
Bedford Institute of Oceanography, Dartmouth, Nova Scotia

and H.L. Windom,
Skidaway Institute of Oceanography, Savannah, Georgia

## Abstract

This report presents the results of the sampling of sea water and the analyses for trace metals in sea water carried out at the Bermuda Biological Station for Research, Inc. in January 1980. This exercise was designed to evaluate the use of different sampling bottles and hydrowires for collecting contamination-free sea-water samples for trace-metal analysis. The evaluation is based on the results of analyses of replicate samples by a number of participating laboratories. An ancillary experiment was also conducted to evaluate the effect of water-storage time in different samplers on trace-metal concentrations.

The experiment reveals better agreement between experienced marine laboratories for several metals than has been the case in previous intercalibrations. Furthermore, this agreement occurs at values very close to contemporary estimates for the trace-metal concentrations of deep ocean waters. The experiment also shows that differences in results derived through the use of different sampling bottles and hydrowires are not as large as has been claimed in the past. Nevertheless, in the case of some metals, notably copper and zinc, pronounced and significant differences do occur between sea-water samples collected with different sampling bottles and hydrowires.

## Introduction

The Pilot Project, in respect of trace, or heavy metals comprises four different activities :

(i) intercalibration of standard solutions,
(ii) intercalibration of low-level ambient sea-water solutions ;
(iii) assessment of sampling procedures for sea-water intended for trace-metal analysis ; and
(iv) training and providing exercises.

These exercises are to be devised and supervised by the IOC Group of Experts on Methods, Standards and Intercalibration (GEMSI). GEMSI has already listed particular metals for which oceanic baselines need to be determined and which correspondingly deserve attention within the Pilot Project programme.

These metals are :

Priority 1 : mercury, cadmium, lead, copper, zinc, nickel, selenium, cobalt ;
Priority 2 : chromium, arsenic, manganese, iron, tin, molybdenum, vanadium.

Elements in the first priority list are believed to be environmentally important either because they are mobilized by human activity on a scale comparable with natural fluxes, or they are believed to be intimately involved in biological processes in the open ocean. Elements in the second priority list are those for which baselines need to be established, although their role in biological processes and their rate of release to the oceans by man is probably less important. However, since 1977, when this list was prepared, our knowledge of the marine geochemistry of metals has improved markedly and it may accordingly require some revision.

The intercalibration activities of the International Council for the Exploration of the Sea (ICES) with regard to trace metals in seawater, which have taken place during the last five years, have provided valuable background information that deserves consideration in the context of planned Pilot Project activities. ICES has conducted the following series of intercalibrations for metals in seawater

1. High-level standards intercalibration (Jones 1976)
2. Intercalibration for mercury in sea water (Olafsson 1976)
3. Low-level intercalibration for trace metals and sea water using North Sea samples (Jones 1977)
4. Low-level intercalibration for trace metals in sea water using Scotian Shelf samples (Bewers *et al* 1979)
5. Intercalibration for cadmium in sea water (Thibaud 1980)
6. Intercalibration for mercury in sea water (Olafsson 1980)

These various intercalibrations illustrate a significant improvement in the sensitivity, precision and comparability of such analyses over the last few years. However, no previous international intercalibrations have tackled the problems and distortions arising from sampling procedures. ICES has long planned to conduct a sampling-method intercomparison once the assessment and refinement of analytical procedures had been completed and this experiment is now in its final planning stages (Bewers *et al* 1978, 1980 ; ICES 1980).

By early 1979 GEMSI had concluded that it would be possible to conduct a sampling intercalibration to examine the differences between different methods for the collection of sea-water samples for trace-metal analysis. The Core Group of GEMSI had, at a meeting in Monaco (March 1979), devised a basic experimental approach, in concept very similar to the designs of ICES. At this same time both IOC and ICES gained considerable benefit from the results of a survey of sampling methods conducted by Prof. H.L. Windom (Windom, 1979). All the ingredients necessary for the preparation of a detailed experimental design were thus available by May 1979 after the ICES Marine Chemistry Working Group meeting. It is to the credit of IOC and GEMSI that efforts were made to involve ICES in the projected sampling intercalibration, but for various administrative reasons ICES was unable to involve itself formally in the subsequent IOC/WMO/UNEP Pilot Project Intercalibration. One item of difference between the objectives of IOC and ICES should, however, be stressed in this connection. The application of IOC's experience, once developed, will be for the examination of deep-ocean metal levels whereas the area of primary interest of ICES is the continental shelf and coastal regions of the North Atlantic. Nevertheless, we believe that the IOC intercalibration reported here will provide benefits to all such organizations having interests in the assessment of oceanic pollution by metals.

This then is the background to the design and organization of the IOC/WMO/UNEP Intercalibration exercise that was held in Bermuda in January 1980. The final plans for the trace-metal component of this exercise were made during the second session of GEMSI in Bermuda, 8-10 October 1979. Further details concerning the trace-metal intercalibration-exercise and other activities conducted during the Workshop may be found in IOC Workshop Report No. 25 (1980).

## I.1 Experimental Objectives and Design

The objectives of the experiment were to assess the intercomparability of the most commonly used procedures for the collection of sea-water samples for trace-metal analysis. Windom's (1979) survey of sampling methods had revealed that the most commonly used approach was the deployment of sampling bottles on hydrowires, but various combinations of these were preferred. The extent to which non-commercial bottle designs and rosette systems were being used in deep-ocean environments was relatively limited. The most common sampling bottle types were General Oceanics NISKIN and GO-FLO samplers and Hydro-Bios samplers, whilst the most widely used hydrowires were kevlar, nylon and steel. Thus, in the case of sampling bottles, the choices for intercomparison are fairly clear, since the three types specified cover the vast majority of marine trace-metal sampling methods. Nevertheless, it was clear from Windom's survey that individual laboratories often made modifications to the commercial bottles in order to improve their performance, especially with respect to minimizing contamination. Thus we believed it desirable to include assessment of such modifications in the experimental design, if possible, since it was intended to assess the bottles' **potential** for trace-metal sample collection rather than their unmodified design characteristics.

In the case of hydrowires, the picture was not quite so unambiguous. Current research activities involving trace-metal determinations in the ocean are carried out using a wide variety of hydrowire materials. In making choices of the few hydrowires that should be intercompared in the IOC/WMO/UNEP intercalibration exercise, special attention was given to hydrowires currently used for the acquisition of deep-ocean samples. Thus the choices were : kevlar, for its wide use and the apparent conviction that it was a clean and convenient wire for trace-metal work ; stainless steel wire, because several North American (MIT, BIO) institutes still use this wire and most of the more classical data in the literature were collected with it ; and plastic-sheathed (coated) steel wire now increasingly used by European laboratories (Kiel and DHI, Hamburg) as a strong and very clean hydrowire.

Having made a number of choices respecting the sampling equipment to be intercompared, it was then necessary to devise an experimental design that would produce the most efficient and effective comparison of the sampling devices. As a result a design was chosen that maximized symmetry at the penalty of some redundancy. This permitted the design to be abbreviated without loss of integrity should logistical and operational exigencies reduce the shiptime available. This design, shown schematically in Figure I.1, was predicated on the assumption that on-site operations would include only sample collection and preservation. All samples obtained would then be analysed by participants at their own laboratory after the Workshop. Furthermore, the design is based on complete independence of each participant's operations, thereby maintaining the viability of the experiment should individual participants lose samples or be otherwise unable to complete their analyses. The design allows a 2-way classification analysis of variance to be applied to the data using the elements shown in Figure I.2. The degree of final replication (4 samples for each sampling combination) is purposely conservative and allows for the loss or rejection of any individual sample in a set without jeopardizing the design integrity.

The application of the design in any meaningful way requires that all of the samples be collected from a homogeneous body of water. Such homogeneity must extend spatially over the range of sampling depths and temporally over the time required to complete the sample collection for each individual participant. The degree of homogeneity should be better than the precision with which individual samples can be analysed such that all the samples collected for any given participant may safely be assumed to have been collected from water of indentical composition. This in turn means that the viability of the experiment is not assured until the extent of water mass homogeneity has been assessed, but, in the absence of reliable measurements of spatial and temporal homogeneity, this is the

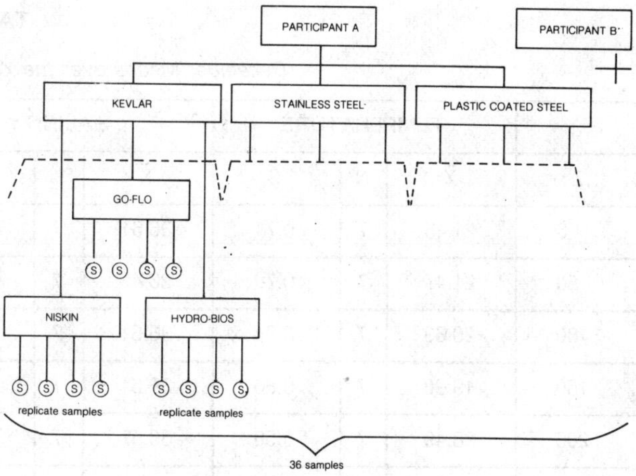


**Figure I.1 Experimental Design**

only option available. Such homogeneity checks should, of course, always be a feature of such experiments to ensure that one's projections of historical data to the sampling scale are correct and that no abnormalities have occurred during the course of the experiment.

For this experiment the location chosen for the sampling activities was ocean station 'S' often referred to as PANULIRUS STATION near Bermuda (Figure 0.1). This location was chosen for a variety of reasons both scientific and logistical. From the scientific standpoint Panulirus station is one of the few ocean stations for which a long and continuous record of physical oceanographic measurements exists for the entire water column to a depth of 2500 metres. Furthermore, the data collected at this station since 1953 attest to the temporal and spatial uniformity of physical conditions in the lower part of the water column (Pocklington 1972a, b). It therefore looked like a most suitable location in which water in a fairly confined intermediate depth range could be assumed to be homogeneous in major (and thereby trace) properties over a period of approximately one month. Such conditions were suitable for the conduct of the trace-metal segment of the IOC/WMO/UNEP intercalibration exercise, and the location was ideal for various logistical reasons, so it was therefore decided to carry out the sampling exercises there. Tables I.1, I.2 and I.3 show the mean and standard deviations of temperature, salinity, $\sigma_t$, and oxygen concentrations for December, January and February over the years 1965-69 after Pocklington (1972a). In order to be well away from both surface variability and bottom influences it was considered suitable to collect intercalibration samples in as confined a depth interval as possible in the range 1400 to 1750 metres but to make provisions to check prevailing conditions and metal distribution homogeneity during the course of the sampling operations.

| | | SAMPLING BOTTLE TYPE | | |
|---|---|---|---|---|
| | | GO-FLO | NISKIN | HYDRO-BIOS |
| HYDRO WIRE TYPE | KEVLAR | 4 Samples | 4 Samples | 4 Samples |
| | STAINLESS STEEL | 4 Samples | 4 Samples | 4 Samples |
| | PLASTIC COATED STEEL | 4 Samples | 4 Samples | 4 Samples |

**Figure I.2 Experimental design matrix elements and sampling density for each participant**

## TABLE I .1

*December Means over the Years 1965, 1966, 1967, 1968, 1969*

| | TEMPERATURE | | | SALINITY | | | SIGMA-T | | | OXYGEN | | |
|---|---|---|---|---|---|---|---|---|---|---|---|---|
| Z | $\bar{X}$ | N | S | $\bar{X}$ | N | S | $\bar{X}$ | N | S | $\bar{X}$ | N | S |
| 0 | 21.45 | 7 | 0.71 | 36.61 | 7 | 0.103 | 25.62 | 7 | 0.135 | 5.04 | 7 | 0.12 |
| 50 | 21.47 | 7 | 0.70 | 36.45 | 7 | 0.409 | 25.59 | 7 | 0.153 | 4.98 | 7 | 0.12 |
| 100 | 20.63 | 7 | 0.84 | 36.61 | 7 | 0.093 | 25.83 | 7 | 0.242 | 4.94 | 7 | 0.16 |
| 150 | 19.28 | 7 | 0.80 | 36.61 | 7 | 0.026 | 26.20 | 7 | 0.210 | 4.62 | 7 | 0.15 |
| 200 | 18.49 | 7 | 0.58 | 36.57 | 7 | 0.041 | 26.37 | 7 | 0.128 | 4.70 | 7 | 0.07 |
| 300 | 17.83 | 7 | 0.34 | 36.51 | 7 | 0.051 | 26.48 | 7 | 0.058 | 4.79 | 7 | 0.08 |
| 400 | 17.37 | 7 | 0.37 | 36.43 | 7 | 0.065 | 26.54 | 7 | 0.047 | 4.76 | 7 | 0.14 |
| 600 | 15.08 | 7 | 0.91 | 36.04 | 7 | 0.155 | 26.77 | 7 | 0.086 | 4.22 | 7 | 0.19 |
| 800 | 10.62 | 7 | 0.97 | 35.39 | 7 | 0.126 | 27.17 | 7 | 0.079 | 3.51 | 7 | 0.10 |
| 1000 | 6.72 | 7 | 0.49 | 35.08 | 7 | 0.035 | 27.54 | 7 | 0.049 | 4.32 | 7 | 0.17 |
| 1200 | 5.14 | 7 | 0.23 | 35.04 | 7 | 0.033 | 27.71 | 7 | 0.016 | 5.35 | 7 | 0.14 |
| 1400 | 4.52 | 7 | 0.05 | 35.03 | 7 | 0.019 | 27.78 | 7 | 0.013 | 5.79 | 7 | 0.08 |
| 1750 | 3.98 | 6 | 0.07 | 35.01 | 6 | 0.021 | 27.82 | 6 | 0.010 | 6.01 | 6 | 0.05 |
| 2000 | 3.75 | 6 | 0.04 | 35.01 | 6 | 0.023 | 27.84 | 6 | 0.014 | 6.05 | 6 | 0.05 |
| 2500 | 3.23 | 5 | 0.05 | 34.97 | 5 | 0.018 | 27.86 | 5 | 0.015 | 5.99 | 5 | 0.14 |

## TABLE I .2

*January Means over the Years 1965, 1966, 1967, 1968, 1969*

| | TEMPERATURE | | | SALINITY | | | SIGMA-T | | | OXYGEN | | |
|---|---|---|---|---|---|---|---|---|---|---|---|---|
| Z | $\bar{X}$ | N | S | $\bar{X}$ | N | S | $\bar{X}$ | N | S | $\bar{X}$ | N | S |
| 0 | 19.98 | 7 | 0.66 | 36.62 | 7 | 0.092 | 26.02 | 7 | 0.143 | 5.13 | 7 | 0.13 |
| 50 | 19.91 | 7 | 0.70 | 36.62 | 7 | 0.094 | 26.04 | 7 | 0.171 | 5.00 | 7 | 0.35 |
| 100 | 19.87 | 7 | 0.78 | 36.61 | 7 | 0.092 | 26.04 | 7 | 0.188 | 4.94 | 7 | 0.36 |
| 150 | 18.88 | 7 | 0.27 | 36.60 | 7 | 0.035 | 26.29 | 7 | 0.074 | 4.77 | 7 | 0.24 |
| 200 | 18.36 | 7 | 0.22 | 36.56 | 7 | 0.026 | 26.40 | 7 | 0.056 | 4.76 | 7 | 0.18 |
| 300 | 17.88 | 7 | 0.16 | 36.52 | 7 | 0.020 | 26.49 | 7 | 0.029 | 4.73 | 7 | 0.13 |
| 400 | 17.39 | 7 | 0.17 | 36.45 | 7 | 0.022 | 26.55 | 7 | 0.024 | 4.71 | 7 | 0.11 |
| 600 | 15.06 | 7 | 0.37 | 36.01 | 7 | 0.076 | 26.76 | 7 | 0.034 | 4.08 | 7 | 0.15 |
| 800 | 10.33 | 7 | 0.84 | 35.36 | 7 | 0.098 | 27.19 | 7 | 0.076 | 3.45 | 7 | 0.07 |
| 1000 | 6.61 | 7 | 0.43 | 35.06 | 7 | 0.030 | 27.54 | 7 | 0.046 | 4.39 | 7 | 0.25 |
| 1200 | 5.08 | 7 | 0.16 | 35.04 | 7 | 0.022 | 27.72 | 7 | 0.009 | 5.39 | 7 | 0.12 |
| 1400 | 4.51 | 7 | 0.08 | 35.03 | 7 | 0.024 | 27.77 | 7 | 0.015 | 5.82 | 7 | 0.13 |
| 1750 | 3.97 | 6 | 0.05 | 35.01 | 7 | 0.020 | 27.81 | 6 | 0.006 | 6.06 | 7 | 0.09 |
| 2000 | 3.72 | 6 | 0.04 | 35.00 | 6 | 0.010 | 27.83 | 6 | 0.008 | 6.08 | 6 | 0.08 |
| 2500 | 3.23 | 5 | 0.03 | 34.97 | 5 | 0.011 | 27.87 | 4 | 0.010 | 6.08 | 6 | 0.07 |

**TABLE I .3**

*February Means over the Years 1965, 1966, 1967, 1968, 1969*

| | TEMPERATURE | | | SALINITY | | | SIGMA-T | | | OXYGEN | | |
|---|---|---|---|---|---|---|---|---|---|---|---|---|
| Z | $\bar{X}$ | N | S | $\bar{X}$ | N | S | $\bar{X}$ | N | S | $\bar{X}$ | N | S |
| 0 | 19.34 | 8 | 0.52 | 36.61 | 8 | 0.090 | 26.18 | 8 | 0.160 | 5.16 | 8 | 0.07 |
| 50 | 19.35 | 8 | 0.50 | 36.57 | 6 | 0.049 | 26.17 | 6 | 0.178 | 5.17 | 8 | 0.06 |
| 100 | 19.32 | 8 | 0.51 | 36.58 | 6 | 0.045 | 26.18 | 6 | 0.178 | 5.14 | 8 | 0.07 |
| 150 | 19.14 | 8 | 0.42 | 36.60 | 6 | 0.032 | 26.23 | 6 | 0.133 | 4.86 | 8 | 0.35 |
| 200 | 18.58 | 8 | 0.22 | 36.59 | 8 | 0.020 | 26.36 | 8 | 0.054 | 4.80 | 8 | 0.26 |
| 300 | 17.85 | 8 | 0.19 | 36.53 | 8 | 0.019 | 26.50 | 8 | 0.037 | 4.70 | 8 | 0.13 |
| 400 | 17.39 | 8 | 0.14 | 36.46 | 8 | 0.031 | 26.56 | 8 | 0.020 | 4.64 | 8 | 0.37 |
| 600 | 15.17 | 8 | 0.33 | 36.04 | 8 | 0.056 | 26.75 | 8 | 0.034 | 4.03 | 8 | 0.26 |
| 800 | 10.60 | 7 | 0.36 | 35.38 | 7 | 0.051 | 27.16 | 7 | 0.037 | 3.46 | 7 | 0.07 |
| 1000 | 6.56 | 7 | 0.35 | 35.07 | 7 | 0.028 | 27.56 | 7 | 0.031 | 4.41 | 7 | 0.19 |
| 1200 | 5.12 | 7 | 0.12 | 35.06 | 7 | 0.018 | 27.73 | 7 | 0.024 | 5.39 | 7 | 0.11 |
| 1400 | 4.52 | 7 | 0.12 | 35.04 | 7 | 0.017 | 27.78 | 7 | 0.020 | 5.84 | 7 | 0.09 |
| 1750 | 3.99 | 7 | 0.05 | 35.01 | 7 | 0.011 | 27.82 | 7 | 0.009 | 6.05 | 7 | 0.09 |
| 2000 | 3.77 | 7 | 0.08 | 35.01 | 7 | 0.015 | 27.84 | 7 | 0.013 | 6.07 | 7 | 0.07 |
| 2500 | 3.28 | 4 | 0.04 | 34.99 | 6 | 0.015 | 27.87 | 4 | 0.006 | 6.04 | 6 | 0.05 |

These plans were formulated during the Second Session of GEMSI in Bermuda in October 1979. As will be seen, operational difficulties forced alterations to the design and logistical aspects of theses plans but the overall approach and objectives remained the same.

## I.2 Experimental Procedures and Logistics

Once the decision had been made to proceed with the exercise in Bermuda in January 1980, potential participants were contacted and asked to bring to Bermuda a minimum of 36 precleaned sample bottles and sufficient quantities of preserving fluids (acids) to enable them to collect 36 samples and return them to their laboratories for subsequent analysis.

It was then necessary to determine the equipment needed for the sampling activities and to ensure that it was assembled at the Bermuda Biological Station. Thanks to the enthusiasm and diligence of several participants it was possible to obtain the following equipment from the sources specified.

| | |
|---|---|
| 1300 m Kevlar hydrowire | Skidaway Institute of Oceanography |
| 3000 m Type 302 stainless-steel hydrowire | Bedford Institute of Oceanography |
| 3000 m Plastic-coated steel hydrowire | Deutsches Hydrographisches Institut |
| 12 litre modified GO-FLO samplers | Bedford Institute of Oceanography |
| 10 litre ex-works GO-FLO samplers | NOAA Atlantic Oceanographic and Meteorological Laboratory in Miami |
| 1.7 litre Hydro-Bios samplers | Deutsches Hydrographisches Institut |
| 12 litre modified Niskin samplers | Bedford Institute of Oceanography |
| Plastic-encased hydro-weight Messengers | Bedford Institute of Oceanography, Deutsches Hydrographisches Institut |

**Meter Block**

R.V. GEORGE B. KELEZ

The oceanographic vessel made available by the US National Oceanographic and Atmospheric Administration (NOAA) was also equipped with many devices and features that were essential to the effective conduct of the experiment. Paramount among these was the clean-van facility maintained by NOAA Miami.

Although not required by the experimental design, the Bermuda Biological Station made available a shore-based clean room (Fig. I.3) equipped with a positive pressure HEPA-filtered air supply, and this room was utilized for precleaning of sampling bottles, sample containers and other gear when necessary. It also enabled participants to carry out some preconcentration, separation and analytical work on site. In particular it enabled a rapid check to be made on water-mass homogeneity, which was most valuable.

For this purpose Varian and Perkin Elmer Flameless Atomic Absorption Spectrophotometers were available (Fig. I.4 and I.5).

Some additional details of the sampling gear are warranted so that the results of the inter-comparison may be thoroughly appreciated.

Unmodified GO-FLO Samplers. These devices were supplied from stock with internal teflon coatings by General Oceanics. No subsequent modifications were made.

Modified GO-FLO Samplers. These devices were supplied from stock with internal teflon coatings by General Oceanics. Subsequent modifications included replacement of all 'O' rings with silcone 'O' rings, replacement of the drain cock with a solid teflon stopcock tapped

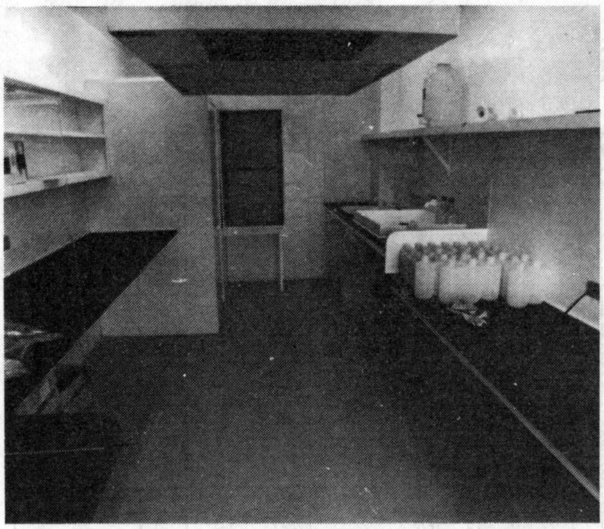

**Figure I.3 Clean Room of the Bermudes Biological Station**

into the body of the bottle and replacement of the original air vent by one made of solid teflon. The bottles were acid washed and rinsed with Super Q water prior to deployment.

Modified Niskin Samplers. These devices were supplied by General Oceanics from stock. They were not teflon coated. These bottles were modified by replacing the internal spring with one made of silicone tubing and replacing all 'O' rings with silicone rubber 'O' rings. The bottles were acid-washed and rinsed with Super Q water prior to deployment.

Kevlar Hydrowire was 1/4'' diameter Kevlar strands covered with a braided nylon sheath.

Stainless Steel Hydrowire was 5/16'' — diameter 3/19 construction unlubricated type 302 stainless steel.

Plastic-coated steel hydrowire was 5/16'' — diameter galvanized-steel wire covered with an approximately 1/16'' wall thickness PVC sheath.

Messengers for the Niskin and GO-FLO bottles were constructed of weighted teflon.

Hydroweights were constructed by pouring concrete into a garbage can of about 4-litre capacity.

All other devices were to our best knowledge of ex-works specifications.

No special precautions were taken with the handling of either the stainless-steel or plastic-coated hydrowires. However, Kevlar wire was mounted on a small winch with relatively clean metal guides. In all cases the meter block was of metal construction but the wheel was kept clean and bright.

## I.3 Shipboard Operations

The sampling operations were carried out during two periods of shiptime — the first from 1230 16 January to 1100 19 January and the second from 1600 20 January to 0100 21 January. At the commencement of the shipboard sampling operations it was still intended to use the experimental design previously formulated. The intended first day operations therefore consisted of collecting all the participants' samples from the three types of sampling bottle on a simple hydrowire (Kevlar). The rather severe weather and sea conditions and the malfunction of the Hydro-Bios bottles, however, forced a change in plans, and between the two periods of shiptime the experimental design was modified by reducing the redundancy in the original design at the cost of the original design's symmetry. A comparison of modified and unmodified GO-FLO samplers was also introduced to the design. Figure 4 shows the elements of the actual experimental design used for the collection of all participants' samples.

Owing to the restricted volume of the Hydro-Bios samplers it was not possible for most participants to obtain replicate samples from a single bottle of this type. Thus in most cases single samples have been drawn from individual Hydro-Bios bottles as compared to quadruplicates from the

**Figure I.4 Varian Flameless Atomic Absorbtion Spectrophotometer**

Niskin and GO-FLO samplers. This operational difference can easily be dealt with within the basic experimental design but any examination of the replication within sampling bottles will need to consider the additional (bottle-to-bottle) source of variance in the case of the Hydro-Bios samplers.

A further problem, the limited length of Kevlar hydrowire available and the capacity of the winch upon which it was loaded, forced us to reduce the depth from which samples were collected. The original plans called for the collection of samples from depths between 1500 and 1750 metres but this had to be changed to 1150-1250 metres depth with a consequent reduction in vertical homogeneity, especially in view of the proximity of the oxygen minimum at 800 metres. This made it even more important to assess the vertical and temporal homogeneity of the metal distributions by collecting samples from the spatial and temporal extremities of the sampling envelope for analysis by individual participants. Such samples were collected. The set from the first period of ship operations were subsequently partly analysed by the University of Delaware participants at the Bermuda Biological Station and the set from the second day's operations by the participants from Skidaway Institute of Oceanography and the Bedford Institute of Oceanography.

The shipboard operations for the two sampling periods are given in Tables I.4 and I.5. The scientific party entrusted with the sampling operations on board the R.V. George B. Kelez was carefully selected on the basis of experience in shipboard sampling operations and familiarity with the types of hydrowire and sampling bottles to be intercompared. The shipboard party was split into two watches which were led on both days by R. Smith of the

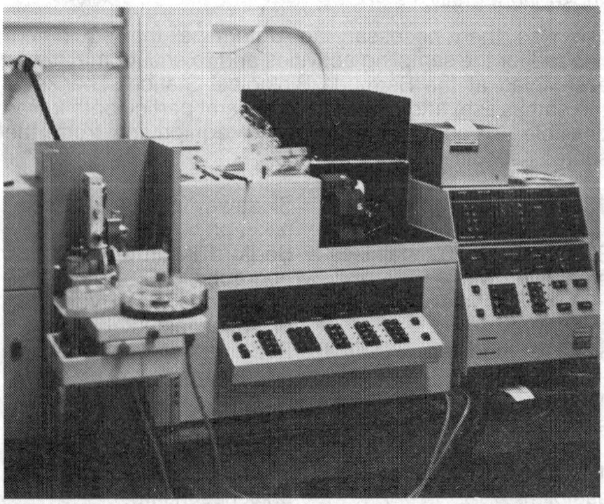

**Figure I.5 Perkin Elmer Flanceless Atomic Absorbtion Spectrophotometer**

## Table I.4

*Shipboard Sampling Operations*
*First Period 8001161230Z — 8001191100Z*

| Cast | Hydrowire type | Sampling device type | Nominal depth m | Identifier No | Samples Collected |
|---|---|---|---|---|---|
| 1-1 | Kevlar | Hydro-Bios | 1175 | 800503 | Homogeneity only |
| | | Niskin | 1183 | 800504 | KORDI X6 NRL X4 |
| | | mod GO-FLO | 1191 | 800505 | KORDI X6 NRL X4 |
| | | Hydro-Bios | 1199 | 800506 | KORDI X1 NRL X1 |
| | | Hydro-Bios | 1207 | 800507 | mistrip |
| | | Hydro-Bios | 1215 | 800508 | KORDI X1 NRL X1 |
| | | Hydro-Bios | 1223 | 800509 | KORDI X1 NRL X1 |
| | | Hydro-Bios | 1231 | 800510 | KORDI X2 |
| | | Hydro-Bios | 1239 | 800511 | Homogeneity only |
| | | Nansen | 1247[1] | 800512 | Salinity only |
| 1-2 | Kevlar | Niskin | 1175 | 800513 | MRI X4 X4 |
| | | mod GO-FLO | 1183 | 800514 | MRI X4 X4 |
| | | Hydro-Bios | 1191 | 800515 | Homogeneity only |
| | | Hydro-Bios | 1199 | 800516 | MRI X1 X2 |
| | | Hydro-Bios | 1207 | 800517 | mistrip |
| | | Hydro-Bios | 1215 | 800518 | MRI X1 X1 |
| | | Hydro-Bios | 1223 | 800519 | MRI X1 X1 |
| | | Hydro-Bios | 1231 | 800520 | MRI X1 X1 |
| | | Hydro-Bios | 1239 | 800521 | Homogeneity only |
| | | Nansen | 1247[2] | 800522 | Salinity only |
| 1-3 | Kevlar | mod GO-FLO | 1217 | 800527 | NISR X4 DHI X4 USM X4 UDEL X3 |
| | | mod GO-FLO | 1225 | 800526 | NISR X4 DHI X4 USM X3 UDEL X3 |
| | | Niskin | 1231 | 800525 | NISR X4 DHI X4 USM X6 UDEL X3 |
| | | Nansen | 1239[3] | 800524 | Salinity only |
| | | Nansen | 1247[4] | 800523 | Salinity only |
| 1-4 | Kevlar | mod GO-FLO | 1215 | 800532 | BIO X4 SKID X4 ifM X4 |
| | | mod GO-FLO | 1223 | 800531 | BIO X3 SKID X4 IfM X2 |
| | | Niskin | 1231 | 800530 | BIO X4 SKID X4 Ifm X4 |
| | | Nansen | 1239[5] | 800529 | Salinity only |
| | | Nansen | 1247 | 800528 | Salinity only |
| 1-5 | Kevlar | mod GO-FLO | 1215 | 800537 | NOAA X4 X4 TAMU X4 NIES X2 |
| | | mod GO-FLO | 1223 | 800536 | NOAA X4 X4 TAMU X4 NIES X2 |
| | | Niskin | 1231 | 800535 | NOAA X4 X4 TAMU X4 NIES X2 |
| | | Nansen | 1239[6] | 800534 | Salinity only |
| | | Nansen | 1247[7] | 800533 | Salinity only |
| 1-6 | Kevlar | mod GO-FLO | 1215 | 800542 | SAG X1 X1 X1 |
| | | mod GO-FLO | 1223 | 800541 | SAG X1 X1 X1 |
| | | Niskin | 1231 | 800540 | SAG X1 X1 X1 |
| | | Nansen | 1239[8] | 800539 | Salinity only |
| | | Nansen | 1247 | 800538 | Salinity only |
| 1-7 | Stainless steel | mod GO-FLO | 1207 | 800543 | MRI X4 X4 DHI X4 USM X4 |
| | | mod GO-FLO | 1215 | 800544 | NRL X4 NISR X4 KORDI X3 |
| | | exw GO-FLO | 1223 | 800545 | MRI X4 X4 DHI X4 USM X4 |
| | | exw GO-FLO | 1231 | 800546 | NRL X4 NISR X4 KORDI X3 |
| | | Nansen | 1239[9] | 800547 | Salinity only |
| | | Nansen | 1247[10] | 800548 | Salinity only |

*1. Corrected depth = 1277 metres Salinity = 35.374 per mil*
*2. Nansen bottle mistripped*
*3. Salinity = 35.086 per mil*
*4. Corrected depth = 1175 metres*
*5. Corrected depth = — metres Salinity = 35.070 per mil*
*6. Salinity = 35.060 per mil*
*7. Corrected depth 1220 metres*
*8. Corrected depth = 1144 metres*
*9. Corrected depth = 1158 metres*
*10. Salinity = 35.072 per mil.*

## Table I.4 (continued)

| Cast | Hydrowire type | Sampling device type | Nominal depth m | Identifier No | Samples Collected |
|---|---|---|---|---|---|
| 1-8 | Stainless Steel | mod GO-FLO | 1207 | 800549 | B10 X4 SKID X4 |
| | | mod GO-FLO | 1215 | 800550 | UDEL X3 |
| | | exw GO-FLO | 1223 | 800551 | BIO X4 SKID X4 |
| | | exw GO-FLO | 1231 | 800552 | UDEL X4 |
| | | Nansen | 1239[1] | 800553 | Salinity only |
| | | Nansen | 1247 | 800554 | Salinity only |
| 1-9 | Stainless Steel | mod GO-FLO | 1207 | 800555 | TAMU X4 NIES X3 |
| | | mod GO-FLO | 1215 | 800556 | IfM X4 NOAA X4 X4 |
| | | exw GO-FLO | 1223 | 800557 | TAMU X4 NIES X3 |
| | | exw GO-FLO | 1231 | 800558 | IfM X4 NOAA X4 X4 |
| | | Nansen | 1239[2] | 800559 | Salinity only |
| | | Nansen | 1247[3] | 800560 | Salinity only |

*1. Corrected depth 1131 metres Salinity = — per mil 2. Corrected depth = 1183 metres 3. Salinity = 35.067 per mil.*

### *Key to Table I.4*

| | |
|---|---|
| mod GO-FLO | *Modified General Oceanics GO-FLO bottle* |
| exw GO-FLO | *Ex-works General Oceanics GO-FLO bottle* |
| BIO | *Bedford Institute of Oceanography, Dartmouth* |
| DHI | *Deutsches Hydrographisches Institut, Hamburg* |
| IfM | *Institut fur Meereskunde, Rostock* |
| KORDI | *Korea Ocean Research and Development Inst.* |
| MRI | *Marine Research Institute, Reykjavik* |
| NIES | *National Institute for Environmental Research* |
| NISR | *Netherlands Institute for Sea Research, Texel* |
| NOAA | *NOAA Laboratory, Miami* |
| NRL | *Naval Research Laboratory, Washington* |
| SAG | *Sagami Chemical Research Centre* |
| SKID | *Skidaway Institute of Oceanography, Savannah* |
| TAMU | *Texas A&M University* |
| UDEL | *University of Delaware* |
| USM | *Universiti Sains Malaysia, Penang* |
| LAB X2 X4 | *Format indicates that the same participant drew different types (volumes) of samples. (Two of one type and four of another)* |

## Table I.5

*Shipboard Sampling Operations*
*Second Period 8001201600Z — 8001210100Z*

| Cast | Hydrowire type | Sampling device type | Nominal depth m | Identifier No | Samples Collected |
|---|---|---|---|---|---|
| 2-1 | PCS | mod GO-FLO | 1167 | 800570 | B10 X3 SKID X3 |
| | | Hydro-Bios | 1175 | 800571 | B10 X1 |
| | | Hydro-Bios | 1183 | 800572 | B10 X1 |
| | | Hydro-Bios | 1191 | 800573 | B10 X1 |
| | | Hydro-Bios | 1199 | 800574 | B10 X1 |
| | | Hydro-Bios | 1207 | 800575 | IfM X1 |
| | | Hydro-Bios | 1215 | 800576 | IfM X1 |
| | | Hydro-Bios | 1223 | 800577 | IfM X1 SKID X3 |
| | | Hydro-Bios | 1231 | 800578 | IfM X1 SKID X3 |
| | | mod GO-FLO | 1239 | 800579 | BIO X3 SKID X3<br>IfM X3 |
| | | Nansen | 1247[1] | 800580 | Salinity only |
| 2-2 | PCS | mod GO-FLO | 1167 | 800581 | MRI X4 X4 |
| | | Hydro-Bios | 1175 | 800582 | MRI X1 X1 |
| | | Hydro-Bios | 1183 | 800583 | MRI X1 X1 |
| | | Hydro-Bios | 1191 | 800584 | MRI X1 X1 |
| | | Hydro-Bios | 1199 | 800585 | MRI X1 X1 |
| | | Hydro-Bios | 1207 | 800586 | SAG X1 |
| | | Hydro-Bios | 1215 | 800587 | SAG X1 |
| | | Hydro-Bios | 1223 | 800588 | NOAA X1 X2 |
| | | Hydro-Bios | 1231 | 800589 | SAG X1<br>NOAA X2 |
| | | mod GO-FLO | 1239 | 800590 | SAG X3<br>NOAA X1 X4 |
| | | Nansen | 1247[2] | 800591 | Salinity only |
| 2-3 | PCS | mod GO-FLO | 1167 | 800592 | UDEL X3 NRL X4 |
| | | Hydro-Bios | 1175 | 800593 | UDEL X1 NRL X1 |
| | | Hydro-Bios | 1183 | 800594 | UDEL X1 NRL X1 |
| | | Hydro-Bios | 1191 | 800595 | UDEL X1 NRL X1 |

*1. Corrected depth = 1160 metres Salinity = 35.389 per mil 2. Corrected depth = 1165 metres Salinity = 35.322 per mil*

**Table I.5 (continued)**

*Shipboard Sampling Operations*

| Cast | Hydrowire type | Sampling device type | Nominal depth m | Identifier No | Samples Collected |
|---|---|---|---|---|---|
| 2-3 (cont.) | PCS | Hydro-Bios | 1199 | 800596 | NISR X1 NRL X1 |
| | | Hydro-Bios | 1207 | 800597 | NISR X1 DHI X1 |
| | | Hydro-Bios | 1215 | 800598 | NISR X1 DHI X1 |
| | | Hydro-Bios | 1223 | 800599 | Broken |
| | | Hydro-Bios | 1231 | 800600 | NISR X1 DH1 X1 |
| | | mod GO-FLO | 1239 | 800601 | NISR X4 DH1 X4 |
| | | Nansen | 1247[1] | 800602 | Salinity only |
| | | (DHI also received a composite sample of 800597/800598/800600) | | | |
| 2-4 | PCS | mod GO-FLO | 1175 | 800603 | TAMU X4 |
| | | Hydro-Bios | 1183 | 800604 | TAMU X1 |
| | | Hydro-Bios | 1191 | 800605 | TAMU X1 |
| | | Hydro-Bios | 1199 | 800606 | TAMU X1 |
| | | Hydro-Bios | 1207 | 800607 | TAMU X1 USM X1 |
| | | Hydro-Bios | 1215 | 800608 | UCON X1 USM X1 |
| | | Hydro-Bios | 1223 | 800609 | UCON X1 USM X1 |
| | | Hydro-Bios | 1231 | 800610 | UCON X1 USM X1 |
| | | mod GO-FLO | 1239 | 800611 | UCON X4 USM X1 |
| | | Nansen | 1247[2] | 800612 | Salinity only |
| 2-5 | PCS | mod GO-FLO | 1167 | 800613 | KORDI X4 B10 X2 SKID X2 |
| | | Hydro-Bios | 1175 | 800614 | KORDI X1 |
| | | Hydro-Bios | 1183 | 800615 | KORDI X1 |
| | | Hydro-Bios | 1191 | 800616 | KORDI X1 |
| | | Hydro-Bios | 1199 | 800617 | KORDI X1 |
| | | Hydro-Bios | 1207 | 800618 | NIES X1 |
| | | Hydro-Bios | 1215 | 800619 | NIES X1 |
| | | Hydro-Bios | 1223 | — | Broken |
| | | Hydro-Bios | 1231 | 800620 | NIES X1 |
| | | mod GO-FLO | 1239 | 800621 | NIES X2 B10 X3 SKID X2 |
| | | Nansen | 1247[3] | 800622 | Salinity only |

*1. Corrected depth = 1252 metres Salinity = 35.317 per mil*
*2. Corrected depth = 1180 metres Salinity = 35.365 per mil*
*3. Corrected depth = 1208 metres Salinity = 35.307 per mil*

***Key to Table I.5***

| | |
|---|---|
| PCS | *Plastic-coated steel* |
| mod GO-FLO | *Modified General Oceanics GO-FLO bottle* |
| BIO | *Bedford Institute of Oceanography, Dartmouth* |
| DHI | *Deutsches Hydrographisches Institut, Hamburg* |
| IfM | *Institut fur Meereskunde, Rostock* |
| KORDI | *Korea Ocean Research and Development Inst.* |
| MRI | *Marine Research Institute, Reykjavik* |
| NIES | *National Insitute for Environmental Research* |
| NISR | *Netherlands Institute for Sea Research, Texel* |
| NOAA | *NOAA Laboratory, Miami* |
| NRL | *Naval Research Laboratory, Washington* |
| SAG | *Sagami Chemical Research Centre* |
| SKID | *Skidaway Institute of Oceanography, Savannah* |
| TAMU | *Texas A&M University* |
| UCON | *University of Connecticut* |
| UDEL | *University of Delaware* |
| USM | *Universiti Sains Malaysia, Penang* |
| LAB X2 X4 | *Format indicates that the same participant drew different types (volumes) of samples. (Two of one type and four of another)* |

Skidaway ınstitute of Oceanography and P.A. Yeats of the Bedford Institute of Oceanography. This procedure was used to ensure, as far as possible, that the shipboard sampling operations were kept as uniform and invariant as possible, so as to reduce as much as possible any variance associated with sampling procedures. It should, however, again be stressed that the experiment is predominantly based upon *within-laboratory* comparisons and only minor use has been made of comparisons of data between laboratories. It should be recorded that the members of shipboard sampling teams performed a complicated task with precision and alacrity under the most difficult weather conditions. Any success in the trace-metal component of this intercalibration exercise is largely due to their dedication and industry.

## I.4 Results

A list of participants and the results of metal analyses of their samples are attached. Table I.6 depicts the types of data available from participants that can be used for numerical and statistical intercomparisons between samples collected with different sampling bottles and hydrowires. It should be stressed that none of the sample analysts was aware of the correlation between sample identifier numbers and the type of sampler and hydrowire with which the samples were collected. The identities of samples were known only to the scientific co-ordinators, specifically Drs. Andersen and Bewers, and neither divulged such information to the other participants or any of the analysts involved. The experiment has, therefore, been con-

**Table I.6**

*Types of data resulting from participant[1] analyses*

| Participant | Element | | | | | | | | |
|---|---|---|---|---|---|---|---|---|---|
| | Cd | Cu | Fe | Mn | Ni | Pb | Zn | Hg | Others |
| Sagami | ★ | ★ | | | | ★ | ★ | ★ | |
| IfM Rostock | ★ | ★ | | | | | | ★ | |
| U. Delaware | ★ | ★ | | | ★ | | | | |
| NISR Texel | ★[1] | ★ | | | ★ | ★ | ★ | | |
| KORDI Seoul | ★ | ★ | | | ★ | | ★ | | |
| NIES Yatabe | ★ | ★ | ★ | | ★ | | ★ | | Mo, V |
| MRI Reykjavik | ★ | ★ | ★ | ★ | ★ | ★ | | ★ | |
| NRL Washington | ★ | ★ | ★ | | ★ | | ★ | | |
| NOAA Miami | | | | | | | | | |
| Texas A&M U. | ★ | ★ | | | ★ | | | | |
| DHI Hamburg | ★ | ★ | ★ | ★ | ★ | | | | |
| USM Penang | ★ | ★ | ★ | | ★ | ★ | ★ | | Co |
| U. Connecticut | ★ | ★ | | | ★ | | ★ | | |
| Skidaway Inst. | ★ | ★ | | | ★ | | ★ | | |
| BIO Dartmouth | ★ | ★ | ★ | ★ | ★ | | ★ | | |

1. *By two analytical techniques*

ducted 'blind'. The details of sample preservation and storage procedures for each corporate participant are given in Table I.7. These procedures were adhered to in the shipboard sub-sampling operations. The homogeneity check analyses carried out by the participants from the University of Delaware at the Bermuda Biological Station were provided to the organizers before the close of the Workshop.

### Analysis of Results

All the participants data sets were entered into a computer file and analysed with the BREAKDOWN subprogram of the Statistical Package for the Social Sciences (SPSS) (Nie *et al*, 1975). Various individual intercomparisons on each participant's data were carried out to determine the numerical and statistical differences between results for samples collected with different sampling bottles or hydrowires. All such analyses were conducted independently for each corporate participant. The objective of these analyses is to determine the degree to which participants agree in their conclusions about systematic trends or effects associated with different sampling devices. No cross-talk between the data from different laboratories was ever intended or carried out, but, as will be seen, it is possible to make qualitative comparisons betwen the results from different laboratories using common sampling techniques. Tables I.8 to I.17 provide pairs of numerical and statistical intercomparisons for each of the tests between sampling devices for each element for which data have been provided by participants.

A cautionary note should be made with regard to these tables. The number of figures provided in the numerical comparison tables should not be taken to imply that all these digits have significance. During the computer analysis of the data a fixed number of digits was provided to avoid truncation and round-off errors in the processing of the results. In preparing the tables an arbitrary choice has been made with regard to the number of digits presented for a given participant and metal. Throughout we have erred on the side of providing a larger number of digits than the number that are significant. The most appropriate way of deducing the number of significant figures is to use the standard deviation values below the mean values. It is possible, however, to see the number of figures that each participant regarded as significant by examining the data returns from participating laboratories.

Before discussing the results in greater detail it is necessary to comment on the results of the homogeneity checks and their significance to the conclusions that may be drawn from the other data. Seldom does there appear an unambiguous consensus with regard to spatial and temporal homogeneity among the few laboratories that analysed check samples. The conclusions of a given homogeneity check should, however, always be judged in the context of the overall precisions of the laboratory concerned, as measured by the standard deviations of its results in the numerical comparison Tables. The better these precisions are, the greater becomes the laboratory's ability to detect inhomogeneities in the trace-metal distributions during the course of the experiment. Often, it can be concluded that those participants having inferior precisions, with respect to at least one of the laboratories carrying out check-sample analyses, should not be capable of discerning differences between the intercomparison samples that are attributable to water-mass inhomogeneities. In addition, it should be remembered that all of the bottle-to-bottle intercomparisons are based upon the analysis of samples collected from adjacent samplers on an individual cast. It is only in the cases of the wire-to-wire comparisons that temporal homogeneity is essential. In a majority of cases, it can be safely concluded that any spatial and temporal variations in trace-metal concentration, within the depth and temporal boundaries of the experiment, are relatively small and should not invalidate the major consensus conclusions of the experiment.

## Table I.7

*Participants' Instructions for Sample Collection*

| Laboratory | Bottle Type | Filling Procedure | Preservation Procedure |
|---|---|---|---|
| Bedford Institute of Oceanography | 2 litre CPE | Remove Red Flash<br>Drain to waste<br>Shake out drops<br>Fill to shoulder | 5 ml ULTREX HC1 using dispenser provided.<br>Cap and tighten with plastic spanner. |
| Deutsches Hydrographique Institut | 500 ml Teflon | Rinse with 100 ml SW sample. | 1 ml redistilled HC1 with dispenser provided. |
| Institut fur Meereskunde | 500 ml Silica with Poly. Stopper (bagged) | Leave small air gap. | 1 ml conc. $HNO_3$ with pipette provided.<br>Wash tip in $HNO_3$ once before use.<br>Cap and reseal bottles in plastic bags. |
| Korea ORDI | 1 litre Poly. | Rinse once.<br>Fill to line on bottle neck. | 5 ml redistilled $HNO_3$ with dispenser provided.<br>Replace double caps. |
| Marine Research Institute | 500 ml Pyrex (bagged) | Rinse three times.<br>Fill to 400 ml level. | 8 ml $HNO_3$ (MERCK 456) using dispenser provided.<br>Rinse dispenser twice with acid before use.<br>Seal tight and replace in plastic bag. |
| | 1 litre CPE | Rinse three times.<br>Leave air gap. | 5 ml redistilled HC1 using dispenser provided.<br>Rinse dispenser twice with acid before use.<br>Seal tight and replace in plastic bag. |
| National Institute for Environmental Studies | 1 litre Poly. | Rinse once.<br>Leave small air gap. | Bottle contains 1 ml ULTREX $HNO_3$. |
| Naval Research Laboratory | 500 ml CPE | Rinse twice with 10 ml. S.W. sample.<br>Fill to 1 cm below shoulder. | 250 ul conc. $HNO_3$ with Eppendorf pipette provided. Place sample in refrigerator. |
| Netherlands Institute for Sea Research | 1 litre CPE | Rinse twice. Fill to 2 cm below neck. | 2 ml MERCK HC1. |
| NOAA AOML | 1 litre Teflon | | 1 ml ULTREX HNO with pipette provided.<br>250 µl ULTREX $HNO_3$ with pipette provided. |
| Sagami Chemical Research Centre | 1 litre Plastic<br>5 litre Poly.<br>500 ml Glass. | Leave small air gap.<br>Leave small air gap.<br>Leave small air gap. | None<br>None<br>Bottle contains acid. |
| Skidaway Institute of Oceanography | 250 ml Teflon (bagged) | Drain. Rinse twice. | 200 µl NBS $HNO_3$ with Eppendorf provided.<br>Tip must be rinsed with 10 % HC1 before dispensing NBS acid.<br>Replace bottle in plastic bag and seal. |
| Texas A&M University | 1 litre CPE (bagged) | Rinse twice.<br>Fill to shoulder. | 500 µl conc. $HNO_3$ with Eppendorf provided. Use new pipette tip each time.<br>Replace bottle in ziplock bag and reseal. |
| University of Delaware | | Rinse twice.<br>Fill to 800 ml. | 2 ml $HNO_3$ with pipette provided. |
| Universiti Sains Malaysia | 300 ml Glass | Rinse three times. | 500 µl MERCK HC1. |

## Table I.8a

*Cadmium Numerical Comparison*
($\mu g\ l^{-1}$)

| WIRE → <br> BOTTLE → | | PCS HB | PCS GF | SS mod GF | SS exw GF | KEV NIS | KEV GF | PCS ΣGF | SS ΣGF | KEV ΣGF |
|---|---|---|---|---|---|---|---|---|---|---|
| LABORATORY | | | | | | | | | | |
| BIO | m | 0.043 | 0.040 | 0.051 | 0.039 | 0.049 | 0.052 | 0.042 | 0.046 | 0.066 |
| | sd | 0.009 | 0.004 | 0.009 | 0.010 | 0.009 | 0.015 | 0.014 | 0.011 | 0.028 |
| DHI | m | 0.033 | 0.017 | 0.093 | 0.082 | 0.037 | 0.036 | 0.015 | 0.088 | 0.034 |
| | sd | 0.015 | 0.011 | 0.049 | 0.019 | 0.010 | 0.009 | 0.009 | 0.035 | 0.010 |
| IfM | m | 0.070 | 0.068 | 0.065 | 1.123 | 0.486 | 0.407 | 0.043 | 0.594 | 0.496 |
| | sd | 0.077 | 0.052 | 0.033 | 2.029 | 0.380 | 0.412 | 0.044 | 1.490 | 0.561 |
| KORDI | m | 0.035 | 0.050 | 0.060 | 0.045 | | | 0.050 | 0.053 | 0.055 |
| | sd | 0.017 | - | 0.042 | 0.007 | | | - | 0.026 | 0.021 |
| MRI | m | 0.035 | 0.024 | 0.021 | 0.062 | 0.028 | 0.030 | 0.019 | 0.042 | 0.030 |
| | sd | 0.001 | 0.002 | 0.008 | 0.019 | 0.004 | 0.004 | 0.007 | 0.025 | 0.004 |
| NIES | m | 0.036 | 0.022 | 0.028 | 0.032 | 0.040 | 0.037 | 0.022 | 0.030 | 0.037 |
| | sd | 0.006 | 0.007 | 0.005 | 0.005 | 0.004 | 0.005 | 0.007 | 0.006 | 0.004 |
| NRL | m | 1.95 | 2.05 | 1.10 | 1.35 | | | 2.05 | 1.24 | 1.15 |
| | sd | 0.60 | 0.53 | 0.44 | 0.13 | | | 0.53 | 0.30 | 0.72 |
| NISR (AA) | m | 0.021 | 0.025 | 0.016 | 0.020 | 0.017 | 0.014 | 0.025 | 0.018 | 0.016 |
| | sd | 0.007 | 0.010 | 0.005 | 0.006 | 0.004 | 0.004 | 0.010 | 0.005 | 0.007 |
| NISR (ASV) | m | 0.021 | 0.019 | 0.020 | 0.020 | 0.025 | 0.020 | 0.019 | 0.020 | 0.021 |
| | sd | 0.0005 | 0.001 | 0.001 | 0.001 | 0.004 | 0.001 | 0.001 | 0.001 | 0.001 |
| SAG | m | 0.035 | 0.036 | | | 0.825 | 0.041 | 0.036 | | 0.108 |
| | sd | 0.003 | 0.003 | | | 1.097 | 0.016 | 0.003 | | 0.142 |
| SKID | m | 0.023 | 0.046 | 0.046 | 0.052 | 0.039 | 0.033 | 0.041 | 0.048 | 0.041 |
| | sd | 0.003 | 0.006 | 0.024 | 0.006 | 0.004 | 0.006 | 0.006 | 0.018 | 0.010 |
| TAMU | m | 0.036 | 0.037 | 0.029 | 0.038 | 0.037 | 0.047 | 0.037 | 0.034 | 0.043 |
| | sd | 0.003 | 0.001 | 0.003 | — | 0.004 | 0.009 | 0.001 | 0.005 | 0.012 |
| UCON | m | 0.046 | 0.022 | | | | | | | |
| | sd | 0.021 | 0.004 | | | | | | | |
| UDEL | m | 0.020 | 0.023 | 0.023 | 0.036 | 0.023 | 0.026 | 0.023 | 0.029 | 0.027 |
| | sd | 0.004 | 0.005 | 0.001 | 0.002 | 0.001 | 0.006 | 0.005 | 0.007 | 0.006 |
| USM | m | | | 6.9 | 6.1 | 9.1 | 7.1 | 9.1 | 6,5 | 7.6 |
| | sd | | | 1.5 | - | - | 1.7 | - | 1.1 | 1.6 |

## Discussion of Results

The results for each element, and the conclusions drawn from their interpretation, are contained in the following sub-sections. The participating institutions are referred to by their acronyms, which are given in the keys to Tables I. 4 and I. 5.

### Cadmium

The data returns for cadmium are summarized in Table I.8a. The results from two laboratories (NRL and USM) and some of those from a third (IfM) are clearly outlyers and have not been used in drawing the following conclusions. The overall mean for cadmium is 0.035 ± 0.016 $\mu g\ l^{-1}$. This is a fairly narrow distribution for the data from twelve laboratories working independently. The mean value is comparable to the current contemporary estimate for cadmium in deep-ocean waters.

Both SKID and UDEL discerned some inhomogeneity in the water mass used for the intercalibration. On the other hand, B10 concluded that the water was spatially and temporally homogeneous for the second day's operations. However, both SKID and UDEL have slightly better precisions than those of B10, as indicated by the standard deviations of replicated sample analyses. Thus it is possible that inhomogeneities, of an extent comparable with the SKID and UDEL precisions, do occur and could not be detected by B10. It should be stressed that most participants have composite sampling/analytical precisions of 10 $ng\ l^{-1}$ or better. Indeed, the precisions obtained by NISR by the ASV method are remarkable in that they are about 1 $ng\ l^{-1}$.

The statistical analysis of the data, shown in Table I. 8b, reveals some evidence that Hydro-Bios bottles yield higher cadmium values than GO-FLO samplers. Nevertheless, this difference is, on average, 1 $ng\ l^{-1}$ and is therefore of no great significance. In most cases in which participants conclude that the effects of hydrowires are significant, it is Kevlar that yields the highest values and Plastic-coated Steel that yields the lowest. The mean difference between the results from GO-FLO bottles on Kevlar and Plastic-coated Steel hydrowires is 8 $ng\ l^{-1}$. Five of the participants found the difference between modified and unmodifield GO-FLO bottles to be significant, with four of them obtaining higher values for cadmium from unmodified bottles. However, the mean difference between the results from the modified and unmodified bottles is only 4 $ng\ l^{-1}$ which is probably negligible at the mean concentrations reported. A comparison of the bottles using samples of homogeneous, metal-depleted, surface water may reveal the extent of such differences more clearly.

**Table I.8b**

*Cadium Statistical Comparisons*

| BASE COMPARISON | PCS GF/GF | PCS HB/GF | SS exw/modGF | KEV GF/GF | KEV NIS/GF | NON HOMOGY | GF WIRES |
|---|---|---|---|---|---|---|---|
| LABORATORY | | | | | | | |
| BIO | NS | NS | Sig MOD>EXW | 90 | 90 GF>NIS | NS (2) | Sig KEV>SS>PCS |
| DHI | NS | Sig HB>GF | NS | NS | NS | | Sig SS>KEV>PCS |
| IfM | Sig | NS | NS | NS | NS | | NC |
| KORDI | | NS | NS | | | | NS |
| MRI | | Sig >GF | Sig EXW>MOD | | NS | | NS |
| NIES | | Sig HB>GF | Sig EXW>MOD | NS | NS | | Sig KEV>SS>PCS |
| NRL | | NS | NS | | | | Sig PCS>KEV>S |
| NISR (AA) | | NS | NS | NS | NS | | NS |
| NISR (ASV) | | Sig HB>GF | NS | Sig | 90 NIS>GF | | Sig KEV>SS>PCS |
| SAG | | NS | | NS | NS | | NS (KEV ε PCS only) |
| SKID | NS | Sig GF>HB | NS | Sig | NS | Sig (2) | NS |
| TAMU | | NS | Sig EXW>MOD | NS | NS | | 90 KEV>PCS>SS |
| UCON | 90 | Sig HB>GF | | | | | |
| UDEL | | NS | Sig EXW>MOD | NS | NS | Sig (1) | NS |
| USM | | NS | NS | NS | Sig NIS>GF | | Sig PCS>KEV>SS |

## Copper

A visual examination of the copper data shown in Table I.9a suggests a bimodal distribution if the extremes (USM) are excluded. However, copper distributions in the oceans have been relatively intensively studied and there appears to be little dispute that its deep-water concentration is about 0.1 $\mu g\ l^{-1}$. One group of laboratories (BIO, MRI, NISR, SKID, TAMU and UDEL) returned data that yield an overall mean (0.13 = 0.04 $\mu g\ l^{-1}$) that is comparable with this consensus value. The remaining group (DHI, IfM, KORDI, NRL, SAG and UCON) give an overall mean (0.51 ± 0.28 $\mu g\ l^{-1}$) that is considerably larger. In the following discussion of the results of the statistical comparisons we have chosen to use only the data from the former group of laboratories but, as may be seen from Table I.9b, the qualitative aspects of these comparisons are also generally supported by the latter group of laboratories.

Only one of the laboratories analysing homogeneity check samples (SKID) observed any inhomogeneity in the copper distribution during the experiment. Moreover, the significance level associated with this conclusion is reduced to 90 %. It is a little surprising that BIO, which has slightly better and more uniform precisions for its copper analyses, did not observe any significant variations in copper concentrations during the same sequence of shipboard operations. It seems safe to assume that, for most of the participants, no variations in copper concentrations occurred that would be large enough to invalidate the statistical comparisons shown in Table I.9b. The results from BIO and NIES indicate that combined sampling and analytical precisions of about 10 ng $l^{-1}$ can be obtained for copper determinations in seawater based upon replicate sample analyses.

The statistical comparisons shown in Table I.9b indicate that, where differences between Hydro-Bios and GO-FLO bottles are found to be significant, the Hydro-Bios bottles always yield the greater concentration. Furthermore, the mean difference is quite large (0.045 $\mu g\ l^{-1}$) and would suggest strongly that Hydro-Bios bottles contaminate samples with copper. None of the participants was able to detect any significant differences between Niskin and GO-FLO samplers. Simply-modified Niskin bottles seem, therefore, to be capable of collecting samples having comparable integrity to those collected with GO-FLO samplers. Four participants observed significant differences between modified and unmodified GO-FLO samplers. In all such cases the unmodified samplers yielded the higher copper values. The mean difference between the two types of GO-FLO samplers is 0.052 $\mu g\ l^{-1}$ which suggests that the modifications made to ex-works bottles have a pronounced effect in reducing the copper contamination within the

## Table I.9a

*Copper Numerical Comparison*
($\mu g\ l^{-1}$)

| WIRE →<br>BOTTLE → | | PCS<br>HB | PCS<br>GF | SS<br>mod<br>GF | SS<br>exw<br>GF | KEV<br>NIS | KEV<br>GF | PCS<br>ΣGF | SS<br>ΣGF | KEV<br>ΣGF |
|---|---|---|---|---|---|---|---|---|---|---|
| LABORATORY | | | | | | | | | | |
| BIO | m | 0.094 | 0.092 | 0.095 | 0.103 | 0.111 | 0.131 | 0.093 | 0.099 | 0.120 |
| | sd | 0.007 | 0.009 | 0.012 | 0.012 | 0.011 | 0.012 | 0.011 | 0.012 | 0.021 |
| DHI | m | 1.000 | 0.765 | 0.553 | 0.620 | 0.455 | 0.272 | 0.650 | 0.586 | 0.403 |
| | sd | 0.857 | 0.289 | 0.261 | 0.100 | 0.487 | 0.185 | 0.291 | 0.186 | 0.298 |
| IfM | m | 0.437 | 0.180 | 0.211 | 0.205 | 0.447 | 0.550 | 0.233 | 0.208 | 0.455 |
| | sd | 0.347 | 0.084 | 0.067 | 0.034 | 0.540 | 0.317 | 0.081 | 0.051 | 0.314 |
| KORDI | m | 0.533 | 0.435 | 1.25 | 1.065 | | | 0.435 | 1.158 | 1.065 |
| | sd | 0.163 | 0.177 | 0.35 | 0.177 | | | 0.177 | 0.252 | 0.177 |
| MRI | m | 0.188 | 0.063 | 0.064 | 0.142 | 0.101 | 0.072 | 0.101 | 0.103 | 0.072 |
| | sd | 0.108 | 0.003 | 0.004 | 0.010 | 0.049 | 0.012 | 0.059 | 0.043 | 0.012 |
| NIES | m | 0.615 | 0.070 | 0.074 | 0.083 | 0.121 | 0.120 | 0.070 | 0.079 | 0.120 |
| | sd | 0.560 | 0.005 | 0.003 | 0.004 | 0.039 | 0.022 | 0.005 | 0.006 | 0.026 |
| NRL | m | 0.35 | 0.27 | 0.71 | 0.28 | | | 0.27 | 0.50 | 0.32 |
| | sd | 0.29 | 0.12 | 0.59 | 0.23 | | | 0.12 | 0.47 | 0.37 |
| NISR (AA) | m | | | | | | | | | |
| | sd | | | | | | | | | |
| NISR (ASV) | m | 0.155 | 0.045 | 0.163 | 0.278 | 0.133 | 0.160 | 0.045 | 0.220 | 0.140 |
| | sd | 0.076 | 0.006 | 0.044 | 0.059 | 0.030 | 0.037 | 0.006 | 0.078 | 0.038 |
| SAG | m | 0.84 | 0.32 | | | 0.35 | 0.55 | 0.32 | | 0.44 |
| | sd | 0.79 | 0.03 | | | 0.02 | 0.21 | 0.03 | | 0.17 |
| SKID | m | 0.123 | 0.135 | 0.158 | 0.119 | 0.096 | 0.100 | 0.130 | 0.138 | 0.101 |
| | sd | 0.015 | 0.003 | 0.033 | 0.032 | 0.015 | 0.019 | 0.024 | 0.037 | 0.019 |
| TAMU | m | 0.195 | 0.137 | 0.102 | 0.106 | 0.109 | 0.132 | 0.137 | 0.104 | 0.149 |
| | sd | 0.089 | 0.027 | 0.005 | 0.001 | 0.013 | 0.019 | 0.027 | 0.004 | 0.073 |
| UCON | m | 0.590 | 0.172 | | | | | | | |
| | sd | 0.325 | 0.040 | | | | | | | |
| UDEL | m | 0.168 | 0.101 | 0.105 | 0.292 | 0.133 | 0.121 | 0.101 | 0.186 | 0.121 |
| | sd | 0.063 | 0.028 | 0.013 | 0.006 | 0.009 | 0.020 | 0.028 | 0.100 | 0.020 |
| USM | m | 6.5 | 5.5 | 20.6 | 10.3 | 10.3 | 7.6 | 5.5 | 16. | 8.1 |
| | sd | 2.6 | 0.5 | 20.6 | - | - | 2.4 | 0.5 | 15. | 2.5 |

devices. Three of the four participants that observed significant differences between the hydrowires found Plastic-coated Steel to yield the lowest values. The mean difference between the hydrowires yielding the highest and lowest results is quite substantial (0.056 $\mu g\ l^{-1}$), if homogeneity was maintained, suggests that hydrowires can have a profound influence upon the results of copper determinations in sea water.

## Iron

The data from DHI, NRL and USM are clearly outlyers, both from the aspect of the other data returned and contemporary estimates of the iron concentration in intermediate-depth ocean waters, and have been excluded from the following remarks. The overall mean and standard deviation for the remaining data is $0.41 \pm 0.29\ \mu g\ l^{-1}$ which is comparable with current reported values for total iron in deep-ocean waters.

BIO concluded that there existed some inhomogeneity in the iron distribution during the course of the second day's shipboard operations. Unfortunately, no other homogeneity checks were carried out. Of all the elements analysed during the course of this experiment, iron, and perhaps manganese, would be expected to exhibit the greatest spatial inhomogeneity. This is because the intercomparison samples were not filtered before analysis and iron and manganese are fairly major constituents of detrital particulate material in the ocean. Thus variations in the concentration of suspended particulate matter might be expected to result in spatial and temporal inhomogeneities for both of these elements. The combined sampling/analytical precisions of NIES are the most uniform at about 50 - 100 ng $l^{-1}$.

In general, Plastic-coated Steel hydrowire seems to give the lower results for iron, but both the hydrowire and bottle comparisons are inconsistent and we choose not to discuss them further in view of the doubtful homogeneity conditions for iron.

## Lead

Although we have listed the data returns for lead from four participants in Table I.11a, the values are in all cases considerably higher than recent estimates by Dr. C.C. Patterson of the California Institute of Technology for the lead content of deep-ocean waters. Only one laboratory (NISR) has reported values close to recent literature estimates, but it would appear that the samples taken during the intercalibration exercise have been significantly contaminated with lead as is common in such exercises. Il could be that

**Table I.9b**

*Copper Statistical Comparisons*

| BASE COMPARISON | S GF/GF | PCS HB/GF | SS exwGF/modGF | KEV GF/GF | KEV NIS/GF | NON HOMOGY | GF WIRES |
|---|---|---|---|---|---|---|---|
| LABORATORY | | | | | | | |
| BIO | | NS | NS | Sig | NS | NS (2) | Sig<br>KEV>SS>PCS |
| DHI | NS | NS | NS | NS | NS | | NS |
| Ifm | Sig | 90<br>HB>GF | NS | NS | NS | | Sig<br>KEV>PCS>SS |
| KORDI | | NS | NS | | | | Sig<br>SS>KEV>PCS |
| MRI | | 90<br>HB>GF | Sig<br>EXW>MOD | | NS | | NS |
| NIES | | Sig<br>GB>GF | Sig<br>EXW>MOD | NS | NS | | Sig<br>KEV>SS>PCS |
| NRL | | NS | NS | | | | NS |
| NISR (AA) | | | | | | | |
| NISR (ASV) | | Sig<br>HB> | Sig<br>EXW>MOD | NS | NS | | Sig<br>SS>KEV>PCS |
| SAG | | NS | | NS | NS | | NS<br>(KEV ε PCS<br>only) |
| SKID | | NS | NS | Sig | NS | 90 (2) | Sig<br>SS>PCS>EV |
| TAMU | | NS | NS | NS | NS | | NS |
| UCON | Sig | Sig<br>HB>GF | | | | | |
| UDEL | | NS | Sig<br>EXW>MOD | NS | NS | NS(1) | NS |
| USM | | NS | NS | NS | 90<br>NIS>GF | | NS |

the conclusions drawn by NISR with regard to the increased lead concentrations in unmodified GO-FLO and in Niskin bottles relative to modified GO-FLO samplers are valid but, without supporting data from other participants with somewhat better precisions than those associated with the NISR results, we believe that it would be inappropriate to draw firm conclusions from the statistical comparisons shown in Table I.11b.

## Manganese

Only three participants have returned manganese results (see Table I.12a). Until fairly recently it might have been concluded that the results of two of these laboratories (BIO and DHI) were comparable with current estimates of the manganese concentrations in deep-ocean waters. Certainly these results do compare well with the most recent studies of manganese in the North Atlantic Ocean but a recent paper dealing with the Pacific Ocean (Landing and Bruland, 1980) reports lower values for the concentration of manganese in deeper Pacific waters that are similar to the results from MRI. Also the precision for manganese determination by MRI ($\sim$ 5 ng $l^{-1}$) is generally better than the precisions of the other two participants.

BIO concluded that some inhomogeneity existed during the experiment as it did for iron. As noted before, this might be due to variations in the concentration of suspended particulate material. Therefore, although two laboratories (BIO and MRI) concluded that significantly lower manganese values were produced using the Stainless Steel hydrowire, the statistical comparisons in Table I.12b must be treated cautiously.

## Mercury

Unfortunately only three participants (IfM, MRI and SAG) have returned analyses for this element, as shown in Table I.13a. Nevertheless, one of these laboratories (MRI) has a great deal of experience in sea water sampling for mercury and its previous performance in seawater intercalibrations for the element has been exemplary. It therefore seems likely that the results of both MRI and IfM reflect the true concentration of mercury at Panulirus Station. Although both these laboratories agree fairly well quantitatively, only MRI draws significant conclusions from the statistical tests (Table I.13b), namely that decreasing mercury concentrations are derived from sampling bottles in the sequence Niskin « GO-FLO » Hydro-Bios. Since MRI habitually uses Hydro-Bios bottles for its seawater sampling for mercury, the statistical tests confirm the suitability of such a choice.

**Table I.10a**

*Iron Numerical Comparison*
(μg l⁻¹)

| WIRE → BOTTLE → | | PCS HB | PCS GF | SS mod GF | SS exw GF | KEV NIS | KEV GF | PCS ΣGF | SS ΣGF | KEV ΣGF |
|---|---|---|---|---|---|---|---|---|---|---|
| LABORATORY | | | | | | | | | | |
| BIO | m | 0.451 | 0.518 | 1.26 | 0.434 | 0.528 | 0.385 | 0.519 | 0.916 | 0.515 |
| | sd | 0.203 | 0.088 | 0.76 | 0.093 | 0.173 | 0.099 | 0.216 | 0.710 | 0.171 |
| DHI | m | 2.81 | 21.83 | 10.1 | 12.7 | 6.13 | 7.50 | 58.5 | 11.38 | 6.84 |
| | sd | 0.43 | 12.06 | 7.6 | 4.9 | 3.53 | 4.77 | 93.4 | 6.07 | 3.88 |
| IfM | m | | | | | | | | | |
| | sd | | | | | | | | | |
| KORDI | m | | | | | | | | | |
| | sd | | | | | | | | | |
| MRI | m | 0.295 | 0.208 | 0.229 | 0.672 | 0.297 | 0.308 | 0.389 | 0.450 | 0.380 |
| | sd | 0.046 | 0.026 | 0.020 | 0.129 | 0.023 | 0.098 | 0.283 | 0.251 | 0.098 |
| NIES | m | 0.269 | 0.172 | 0.158 | 0.221 | 0.269 | 0.232 | 0.172 | 0.192 | 0.285 |
| | sd | 0.095 | 0.055 | 0.031 | 0.067 | 0.073 | 0.047 | 0.055 | 0.062 | 0.082 |
| NRL | m | 2.83 | 2.48 | 5.60 | 5.78 | | | 2.48 | 5.69 | 4.10 |
| | sd | 1.35 | 1.46 | 1.71 | 1.16 | | | 1.46 | 1.35 | 1.33 |
| NISR (AA) | m | | | | | | | | | |
| | sd | | | | | | | | | |
| NISR (ASV) | m | | | | | | | | | |
| | sd | | | | | | | | | |
| SAG | m | | | | | | | | | |
| | sd | | | | | | | | | |
| SKID | m | | | | | | | | | |
| | sd | | | | | | | | | |
| TAMU | m | | | | | | | | | |
| | sd | | | | | | | | | |
| UCON | m | | | | | | | | | |
| | sd | | | | | | | | | |
| UDEL | m | | | | | | | | | |
| | sd | | | | | | | | | |
| USM | m | 56. | 48. | 39. | 42. | 40. | 53. | 49. | 40. | 43. |
| | sd | 8. | 22. | 8. | 14. | 8. | 5 | 22. | 10. | 15. |

## Molybdenum

Only one laboratory (NIES) has returned analyses of molybdenum (Table I.14a). The results are comparable to previous estimates of the concentration of this metal in normal ocean waters and probably represent the true molybdenum concentration at Panulirus Station. Although most of the statistical tests carried out on the molybdenum data, the results of which are shown in Table I.14b, reveal significant differences between bottles and hydrowires, it would be premature to draw hard conclusions from such a limited data set.

## Nickel

The data from two laboratories (NLR and USM) are clearly outlyers from the rest of the population and have been excluded from further consideration. The overall mean of the remaining data is 0.28 ± 0.12 μg l⁻¹ but it seems evident that the distribution is bimodal. One group of seven laboratories (BIO, DHI, MRI, NIES, NISR, UCON and UDEL) has a mean of 0.209 ± 0.050 μg l⁻¹ while another group of three participants (KORDI, SKID and TAMU) give a mean of 0.416 ± 0.111 μg l⁻¹. This is interesting in view of the divergence of literature reports regarding the concentration of nickel in deep-ocean waters, centering around 0.2 and 0.4 μg l⁻¹. From the results of this experiment, it seems probable that this divergence is a consequence of preservation or analytical techniques, rather than an artifact of sampling method or real oceanic variability. Combined sampling/analytical precisions of about 20 ng l⁻¹ were obtained by several participants.

Neither UDEL nor SKID discerned any inhomogeneity during the course of the first or second day's shipboard operations, respectively. However, BIO, which has generally better precisions than these laboratories, did conclude that inhomogeneities were evident during the second day's operations. It would seem that only NIES and NISR would have detected differences due to inhomogeneity in their intercomparisons since both of these laboratories have slightly better precisions than BIO.

The statistical tests (Table I.15b) reveal an overwhelming consensus that significantly higher concentrations of nickel are derived from Hydro-Bios bottles as compared with GO-FLO samplers. The mean difference between these sampling devices is 0.135 μg l⁻¹ and we conclude that Hydro-Bios bottles can give rise to severe sample contamination for nickel. Although two laboratories (BIO and DHI) observe significant differences between Niskin and GO-FLO samplers, in one case the result is weakened by a conclusion that significant differences also exist betwen GO-FLO

**Table I.10b**
*Iron Statistical Comparisons*

| SE COMPARISON | PCS GF/GF | PCS HB/GF | SS exwGF/modGF | KEV GF/GF | KEV NIS/GF | NON HOMOGY | GF WIRES |
|---|---|---|---|---|---|---|---|
| LABORATORY | | | | | | | |
| BIO | | NS | Sig MOD>EXW | Sig | NS | Sig (2) | Sig SS>PCS>KEV |
| DHI | NS | Sig GF>HB | NS | NS | NS | | NS |
| IfM | | | | | | | |
| KORDI | | | | | | | |
| MRI | | Sig HB>GF | Sig EXW>MOD | | NS | | NS |
| NIES | | Sig HB>GF | Sig EXW>MOD | Sig | NS | | Sig KEV>SS>PCS |
| NRL | | NS | NS | | | | Sig SS>KEV>PCS |
| NISR (AA) | | | | | | | |
| NISR (ASV) | | | | | | | |
| SAG | | | | | | | |
| SKID | | | | | | | |
| TAMU | | | | | | | |
| UCON | | | | | | | |
| UDEL | | | | | | | |
| USM | | NS | NS | NS | NS | | NS |

samplers on the same hydrowire (Kevlar). Most laboratories could discern no differences between Niskin and GO-FLO bottles. Furthermore, the mean difference is only 0.012 $\mu g\ l^{-1}$ which is comparatively small. Two laboratories conclude that unmodified GO-FLO bottles yield higher nickel values than modified bottles, but this observation is counterbalanced by six other laboratories' results which show no such significant differences. There exists a strong majority opinion (only one laboratory dissents) that Plastic-coated Steel hydrowire yields the lowest nickel values, although in only four cases is the difference found to be significant. It might seem prudent to avoid the use of Kevlar and Stainless Steel hydrowires for the collection of seawater samples for nickel analysis, but it should be stressed that the average concentration range between the 'cleanest' and 'dirtiest' wires is only 0.043 $\mu g\ l^{-1}$.

## Vanadium

One laboratory (NIES) determined vanadium in the intercomparison samples (see Table I.16a). The range of values is quite narrow (1.28 to 1.41 $\mu g\ l^{-1}$) and the method used yields a combined sampling/analytical precision of about 0.05 $\mu g\ l^{-1}$. All of the bottle-to-bottle comparisons (Table I.16b) exhibit significant differences between the vanadium values in samples collected with them but, in the absence of other supporting data, we believe it inappropriate to draw hard conclusions from this data set.

## Zinc

The data returned by KORDI, NISR, SAG and USM contain markedly higher zinc levels than either the remaining participants' data or current estimates of the concentration of zinc in deep-ocean waters. These data have therefore been excluded in deriving conclusions from the numerical and statistical comparisons for zinc shown in Tables I.17a and I.17b, respectively. The overall mean for the remaining data is 0.35 ± 0.18 $\mu g\ l^{-1}$ which is comparable with contemporary estimates for the concentration of zinc in intermediate and deep-ocean waters by experienced marine laboratories.

The results of the homogeneity-check sample analyses by BIO and SKID disagree but there is some suggestion that the precision of the SKID analyses is better than that of BIO might explain the differing conclusions. Only BIO and NIES found significant differences between Hydro-Bios and GO-FLO samplers but the NIES conclusion must be considered in the light of the very much higher results from Hydro-Bios bottles than those of the other participants. This may reflect some severe contamination of these particular NIES samples. A better consensus would be required to draw firm conclusions generally about the performance of Hydro-Bios bottles for sea-water sample collection for zinc. In contrast there exists strong evidence that modified GO-FLO bottles yield significantly lower zinc values compared to unmodified bottles. All but one of the laboratories that made this comparison concluded that this difference was significant. The mean difference is 0.13 $\mu g\ l^{-1}$ but it should be noted that it varies considerably among the few participants involved. A majority of the participants also concluded that Niskin bottles give higher zinc values than do GO-FLO bottles. The differences are consistent and give a mean of 0.12 $\mu g\ l^{-1}$. A majority of the participants also concluded that Plastic-coated Steel hydrowire yields the lowest zinc values, and Kevlar hydrowire, the highest. The mean range of zinc concentration between extremes for the three hydrowires is 0.27 $\mu g\ l^{-1}$. The magnitude of this range and the consistency of the conclusions from hydrowire intercomparisons suggest that Plastic-coated Steel hydrowire should be the preferred choice for sea-water sampling for zinc.

## Table I.11a

*Lead Numerical Comparison*
($\mu g\ l^{-1}$)

| WIRE → BOTTLE → | | PCS HB | PCS GF | SS mod GF | SS exw GF | KEV NIS | KEV GF | PCS $\Sigma$GF | SS $\Sigma$GF | KEV $\Sigma$GF |
|---|---|---|---|---|---|---|---|---|---|---|
| LABORATORY | | | | | | | | | | |
| BIO | m | | | | | | | | | |
| | sd | | | | | | | | | |
| DHI | m | | | | | | | | | |
| | sd | | | | | | | | | |
| IfM | m | | | | | | | | | |
| | sd | | | | | | | | | |
| KORDI | m | | | | | | | | | |
| | sd | | | | | | | | | |
| MRI | m | 0.105 | 0.194 | 0.135 | 0.139 | 0.097 | 0.110 | 0.210 | 0.137 | 0.110 |
| | sd | 0.027 | 0.060 | 0.008 | 0.036 | 0.023 | 0.042 | 0.053 | 0.024 | 0.042 |
| NIES | m | | | | | | | | | |
| | sd | | | | | | | | | |
| NRL | m | | | | | | | | | |
| | sd | | | | | | | | | |
| NISR | m | | | | | | | | | |
| (AA) | sd | | | | | | | | | |
| NISR | m | 0.033 | 0.040 | 0.040 | 0.078 | 0.128 | 0.038 | 0.040 | 0.059 | 0.071 |
| (ASV) | sd | 0.026 | - | - | 0.021 | 0.022 | 0.015 | - | 0.024 | 0.041 |
| SAG | m | 0.27 | 0.19 | | | 0.31 | 0.27 | 0.19 | | 0.24 |
| | sd | 0.05 | 0.03 | | | 0.04 | 0.05 | 0.03 | | 0.04 |
| SKID | m | | | | | | | | | |
| | sd | | | | | | | | | |
| TAMU | m | | | | | | | | | |
| | sd | | | | | | | | | |
| UCON | m | | | | | | | | | |
| | sd | | | | | | | | | |
| UDEL | m | | | | | | | | | |
| | sd | | | | | | | | | |
| USM | m | 54. | 56. | 56. | 64. | 60. | 62. | 56. | 59. | 56. |
| | sd | 9. | 10. | 12. | 4. | 3. | - | 10. | 10. | 6. |

## Table I.11b

*Lead Statistical Comparisons*

| BASE COMPARISON | PCS GF/GF | PCS HB/GF | SS exwGF/modGF | KEV GF/GF | KEV NIS/GF | NON HOMOGY | GF WIRES |
|---|---|---|---|---|---|---|---|
| LABORATORY | | | | | | | |
| BIO | | | | | | | |
| DHI | | | | | | | |
| IfM | | | | | | | |
| KORDI | | | | | | | |
| MRI | | Sig GF>HB | NS | | NS | | Sig PCS>SS>KEV |
| NIES | | | | | | | |
| NRL | | | | | | | |
| NISR (AA) | | | | | | | |
| NISR (ASV) | | NS | Sig EXW>MOD | Sig | Sig NIS>GF | | NS |
| SAG | | NS | | NS | NS | | NS (KEV ε PCS only) |
| SKID | | | | | | | |
| TAMU | | | | | | | |
| UCON | | | | | | | |
| UDEL | | | | | | | |
| USM | | NS | NS | Sig | NS | | NS |

**Table I.12a**

*Manganese Numerical Comparison*
($\mu g\ l^{-1}$)

| WIRE → BOTTLE → | | PCS HB | PCS GF | SS mod GF | SS exw GF | KEV NIS | KEV GF | PCS ΣGF | SS ΣGF | KEV ΣGF |
|---|---|---|---|---|---|---|---|---|---|---|
| LABORATORY | | | | | | | | | | |
| BIO | m | 0.047 | 0.091 | 0.046 | 0.042 | 0.053 | 0.052 | 0.062 | 0.044 | 0.052 |
| | sd | 0.015 | 0.045 | 0.018 | 0.012 | 0.008 | 0.003 | 0.032 | 0.014 | 0.005 |
| DHI | m | 0.040 | 0.200 | 0.058 | 0.088 | 0.050 | 0.046 | 0.228 | 0.074 | 0.048 |
| | sd | 0.016 | 0.078 | 0.022 | 0.065 | 0.025 | 0.013 | 0.166 | 0.051 | 0.028 |
| IfM | m | | | | | | | | | |
| | sd | | | | | | | | | |
| KORDI | m | | | | | | | | | |
| | sd | | | | | | | | | |
| MRI | m | 0.009 | 0.008 | 0.007 | 0.009 | 0.012 | 0.023 | 0.013 | 0.008 | 0.023 |
| | sd | 0.002 | 0.002 | 0.002 | 0.004 | 0.009 | 0.005 | 0.008 | 0.003 | 0.005 |
| NIES | m | | | | | | | | | |
| | sd | | | | | | | | | |
| NRL | m | | | | | | | | | |
| | sd | | | | | | | | | |
| NISR (AA) | m | | | | | | | | | |
| | sd | | | | | | | | | |
| NISR (ASV) | m | | | | | | | | | |
| | sd | | | | | | | | | |
| SAG | m | | | | | | | | | |
| | sd | | | | | | | | | |
| SKID | m | | | | | | | | | |
| | sd | | | | | | | | | |
| TAMU | m | | | | | | | | | |
| | sd | | | | | | | | | |
| UCON | m | | | | | | | | | |
| | sd | | | | | | | | | |
| UDEL | m | | | | | | | | | |
| | sd | | | | | | | | | |
| USM | m | | | | | | | | | |
| | sd | | | | | | | | | |

**Table I.12b**

*Manganese Statistical Comparisons*

| BASE COMPARISON | PCS GF/GF | S HB/GF | SS exwGF/modGF | KEV GF/GF | KEV NIS/GF | NON HOMOGY | GF WIRES |
|---|---|---|---|---|---|---|---|
| LABORATORY | | | | | | | |
| BIO | | NS | NS | NS | NS | Sig | 90 PCS>KEV>SS |
| DHI | NS | Sig GF>HB | NS | NS | NS | | Sig PCS>SS>KEV |
| IfM | | | | | | | |
| KORDI | | | | | | | |
| MRI | | NS | NS | | | NS | Sig KEV>PCS>SS |
| NIES | | | | | | | |
| NRL | | | | | | | |
| NISR (AA) | | | | | | | |
| NISR (ASV) | | | | | | | |
| SAG | | | | | | | |
| SKID | | | | | | | |
| TAMU | | | | | | | |
| UCON | | | | | | | |
| UDEL | | | | | | | |
| USM | | | | | | | |

**Table I.13a**

*Mercury Numerical Comparison*
(μg $l^{-1}$)

| WIRE → BOTTLE → | | PCS HB | PCS GF | SS mod GF | SS exw GF | KEV NIS | KEV GF | PCS ΣGF | SS ΣGF | KEV ΣGF |
|---|---|---|---|---|---|---|---|---|---|---|
| LABORATORY | | | | | | | | | | |
| BIO | m | | | | | | | | | |
| | sd | | | | | | | | | |
| DH8 | m | | | | | | | | | |
| | sd | | | | | | | | | |
| IfM | m | 6.1 | 10.4 | 8.5 | 8.2 | 12.8 | 5.4 | 9.5 | 8.4 | 8.1 |
| | sd | 5.8 | 10.1 | 9.1 | 8.1 | 14.0 | 0.3 | 9.2 | 8.3 | 4.7 |
| KORDI | m | | | | | | | | | |
| | sd | | | | | | | | | |
| MRI | m | 4.2 | 6.2 | 3.8 | 10.1 | 6.5 | 4.5 | 6.2 | 6.9 | 4.5 |
| | sd | 1.3 | 4.0 | 1.0 | 5.1 | 0.8 | 1.7 | 4.0 | 4.7 | 1.7 |
| NIES | m | | | | | | | | | |
| | sd | | | | | | | | | |
| NRL | m | | | | | | | | | |
| | sd | | | | | | | | | |
| NISR (AA) | m | | | | | | | | | |
| | sd | | | | | | | | | |
| NISR (ASV) | m | | | | | | | | | |
| | sd | | | | | | | | | |
| SAG | m | 1.1 | | | | 1.0 | 1.1 | | | |
| | sd | 0.1 | | | | 0.1 | 0.1 | | | |
| SKID | m | | | | | | | | | |
| | sd | | | | | | | | | |
| TAMU | m | | | | | | | | | |
| | sd | | | | | | | | | |
| UCON | m | | | | | | | | | |
| | sd | | | | | | | | | |
| UDEL | m | | | | | | | | | |
| | sd | | | | | | | | | |
| USM | m | | | | | | | | | |
| | sd | | | | | | | | | |

**Table I.13b**

*Mercury Statistical Comparisons*

| BASE COMPARISON | PCS GF/GF | PCS HB/GF | SS exwGF/modGF | KEV GF/GF | KEV NIS/GF | KEV BOTTLES | GF WIRES |
|---|---|---|---|---|---|---|---|
| LABORATORY | | | | | | | |
| BIO | | | | | | | |
| DHI | | | | | | | |
| Ifm | NS | NS | NS | NS | NS | | NS |
| KORDI | | | | | | | |
| MRI | | NS | NS | | 90 NIS>GF | Sig NIS>GF> HB | NS |
| NIES | | | | | | | |
| NRL | | | | | | | |
| NISR (AA) | | | | | | | |
| NISR (ASV) | | | | | | | |
| SAG | | | | NS | NS | | |
| SKID | | | | | | | |
| TAMU | | | | | | | |
| UCON | | | | | | | |
| UDEL | | | | | | | |
| USM | | | | | | | |

## Table I.14a

*Molybdenum Numerical Comparison*
($\mu g\ l^{-1}$)

| WIRE → BOTTLE → | | PCS HB | PCS GF | SS mod GF | SS exw GF | KEV NIS | KEV GF | PCS ΣGF | SS ΣGF | KEV ΣGF |
|---|---|---|---|---|---|---|---|---|---|---|
| LABORATORY | | | | | | | | | | |
| BIO | m | | | | | | | | | |
| | sd | | | | | | | | | |
| DHI | m | | | | | | | | | |
| | sd | | | | | | | | | |
| IfM | m | | | | | | | | | |
| | sd | | | | | | | | | |
| KORDI | m | | | | | | | | | |
| | sd | | | | | | | | | |
| MRI | m | | | | | | | | | |
| | sd | | | | | | | | | |
| NIES | m | 7.66 | 8.27 | 7.76 | 8.03 | 7.52 | 8.41 | 8.27 | 7.90 | 8.48 |
| | sd | 0.24 | 0.21 | 0.38 | 0.27 | 1.14 | 0.28 | 0.21 | 0.35 | 0.35 |
| NRL | m | | | | | | | | | |
| | sd | | | | | | | | | |
| NISR (AA) | m | | | | | | | | | |
| | sd | | | | | | | | | |
| NISR (ASV) | m | | | | | | | | | |
| | sd | | | | | | | | | |
| SAG | m | | | | | | | | | |
| | sd | | | | | | | | | |
| SKID | m | | | | | | | | | |
| | sd | | | | | | | | | |
| TAMU | m | | | | | | | | | |
| | sd | | | | | | | | | |
| UCON | m | | | | | | | | | |
| | sd | | | | | | | | | |
| UDEL | m | | | | | | | | | |
| | sd | | | | | | | | | |
| USM | m | | | | | | | | | |
| | sd | | | | | | | | | |

## Table I.14b

*Molybdenum Statistical Comparisons*

| BASE COMPARISON | PCS GF/GF | PCS HB/GF | SS exwGF/modGF | KEV GF/GF | KEV NIS/GF | NON HOMOGY | GF WIRES |
|---|---|---|---|---|---|---|---|
| LABORATORY | | | | | | | |
| BIO | | | | | | | |
| DHI | | | | | | | |
| IfM | | | | | | | |
| KORDI | | | | | | | |
| MRI | | | | | | | |
| NIES | | Sig GF>HB | Sig EXW>MOD | NS | Sig GF>NIS | | Sig KEV>PCS>SS |
| NRL | | | | | | | |
| NISR (AA) | | | | | | | |
| NISR (ASV) | | | | | | | |
| SAG | | | | | | | |
| SKID | | | | | | | |
| TAMU | | | | | | | |
| UCON | | | | | | | |
| UDEL | | | | | | | |
| USM | | | | | | | |

## Table I.15a

*Nickel Numerical Comparison* (µg l$^{-1}$)

| WIRE → BOTTLE → | | PCS HB | PCS GF | SS mod GF | SS exw GF | KEV NIS | KEV GF | PCS ΣGF | SS ΣGF | KEV ΣGF |
|---|---|---|---|---|---|---|---|---|---|---|
| LABORATORY | | | | | | | | | | |
| BIO | m | 0.224 | 0.207 | 0.209 | 0.243 | 0.233 | 0.210 | 0.201 | 0.226 | 0.207 |
| | sd | 0.015 | 0.018 | 0.023 | 0.013 | 0.023 | 0.030 | 0.029 | 0.025 | 0.025 |
| DHI | m | 0.298 | 0.278 | 0.235 | 0.235 | 0.218 | 0.150 | 0.223 | 0.235 | 0.231 |
| | sd | 0.031 | 0.123 | 0.077 | 0.048 | 0.019 | 0.018 | 0.128 | 0.059 | 0.119 |
| IfM | m | | | | | | | | | |
| | sd | | | | | | | | | |
| KORDI | m | 0.18 | 0.47 | 0.47 | 0.35 | | | 0.47 | 0.41 | 0.23 |
| | sd | 0.07 | - | - | 0.17 | | | - | 0.12 | - |
| MRI | m | 0.478 | 0.221 | 0.240 | 0.235 | 0.273 | 0.237 | 0.193 | 0.238 | 0.237 |
| | sd | 0.066 | 0.026 | 0.014 | 0.012 | 0.021 | 0.016 | 0.048 | 0.012 | 0.016 |
| NIES | m | 0.340 | 0.159 | 0.220 | 0.238 | 0.237 | 0.232 | 0.159 | 0.230 | 0.230 |
| | sd | 0.052 | 0.035 | 0.030 | 0.021 | 0.023 | 0.026 | 0.035 | 0.027 | 0.024 |
| NRL | m | 1.93 | 1.63 | 1.60 | 1.75 | | | 1.63 | 1.68 | 1.72 |
| | sd | 0.24 | 0.10 | 0.37 | 0.13 | | | 0.10 | 0.27 | 0.25 |
| NISR (AA) | m | 0.185 | 0.100 | 0.123 | 0.113 | 0.160 | 0.105 | 0.100 | 0.119 | 0.123 |
| | sd | 0.041 | — | 0.005 | 0.012 | 0.054 | 0.010 | - | 0.009 | 0.041 |
| NISR (ASV) | m | | | | | | | | | |
| | sd | | | | | | | | | |
| SAG | m | | | | | | | | | |
| | sd | | | | | | | | | |
| SKID | m | 0.737 | 0.357 | 0.634 | 0.353 | 0.385 | 0.461 | 0.413 | 0.493 | 0.462 |
| | sd | 0.078 | 0.077 | 0.309 | 0.131 | 0.064 | 0.162 | 0.159 | 0.262 | 0.162 |
| TAMU | m | 0.511 | 0.367 | 0.349 | 0.365 | 0.421 | 0.393 | 0.367 | 0.357 | 0.404 |
| | sd | 0.034 | 0.008 | 0.034 | 0.009 | 0.009 | 0.027 | 0.008 | 0.025 | 0.037 |
| UCON | m | 0.230 | 0.165 | | | | | | | |
| | sd | 0.024 | 0.036 | | | | | | | |
| UDEL | m | 0.230 | 0.200 | 0.238 | 0.265 | 0.204 | 0.236 | 0.200 | 0.250 | 0.236 |
| | sd | 0.036 | 0.020 | 0.045 | 0.007 | 0.056 | 0.007 | 0.020 | 0.035 | 0.007 |
| USM | m | 136. | 172. | 145. | 133. | 145. | 127. | 172. | 140. | 124. |
| | sd | 10. | 18. | 30. | 10. | | 31. | 18. | 23. | 21. |

## Table I.15b

*Nickel Statistical Comparisons*

| BASE COMPARISON | PCS GF/GF | PCS HB/GF | SS exwGF/modGF | KEV GF/GF | KEV NIS/GF | NON HOMOGY | GF WIRES |
|---|---|---|---|---|---|---|---|
| LABORATORY | | | | | | | |
| BIO | NS | Sig HB>GF | Sig EXW>MOD | NS | Sig NIS>GF | Sig (2) | Sig SS>KEV>PCS |
| DHI | NS | NS | NS | Sig | Sig NIS>GF | | NS |
| IfM | | | | | | | |
| KORDI | | Sig GF>HB | NS | | | | NS |
| MRI | | Sig HB>GF | NS | | 90 NIS>GF | | Sig SS>KEV>PCS |
| NIES | | Sig HB>GF | Sig EXW>MOD | NS | NS | | Sig KEV>SS>PCS |
| NRL | | 90 HB>GF | NS | | | | NS |
| NISR (AA) | | Sig HB>GF | NS | NS | NS | | NS |
| NISR (ASV) | | | | | | | |
| SAG | | | | | | | |
| SKID | NS | Sig HB>GF | NS | Sig | NS | NS (2) | NS |
| TAMU | | Sig HB>GF | NS | NS | NS | | Sig KEV>PCS>SS |
| UCON | 90 | Sig HB>GF | | | | | |
| UDEL | | NS | NS | NS | NS | NS (1) | 90 SS>KEV>PCS |
| USM | | Sig GF>HB | NS | NS | Sig NIS>GF | | Sig PCS>SS>KEV |

## Table I.16a

*Vanadium Numerical Comparison*
*(µg l⁻¹)*

| WIRE → BOTTLE → | | PCS HB | PCS GF | SS mod GF | SS exw GF | KEV NIS | KEV GF | PCS ΣGF | SS ΣGF | KEV ΣGF |
|---|---|---|---|---|---|---|---|---|---|---|
| LABORATORY | | | | | | | | | | |
| BIO | m | | | | | | | | | |
| | sd | | | | | | | | | |
| DHI | m | | | | | | | | | |
| | sd | | | | | | | | | |
| IfM | m | | | | | | | | | |
| | sd | | | | | | | | | |
| KORDI | m | | | | | | | | | |
| | sd | | | | | | | | | |
| MRI | m | | | | | | | | | |
| | sd | | | | | | | | | |
| NIES | m | 1.28 | 1.39 | 1.41 | 1.37 | 1.33 | 1.39 | 1.39 | 1.38 | 1.36 |
| | sd | 0.02 | 0.06 | 0.03 | 0.04 | 0.10 | 0.05 | 0.06 | 0.04 | 0.28 |
| NRL | m | | | | | | | | | |
| | sd | | | | | | | | | |
| NISR (AA) | m | | | | | | | | | |
| | sd | | | | | | | | | |
| NISR (ASV) | m | | | | | | | | | |
| | sd | | | | | | | | | |
| SAG | m | | | | | | | | | |
| | sd | | | | | | | | | |
| SKID | m | | | | | | | | | |
| | sd | | | | | | | | | |
| TAMU | m | | | | | | | | | |
| | sd | | | | | | | | | |
| UCON | m | | | | | | | | | |
| | sd | | | | | | | | | |
| UDEL | m | | | | | | | | | |
| | sd | | | | | | | | | |
| USM | m | | | | | | | | | |
| | sd | | | | | | | | | |

## Table I.16b

*Vanadium Statistical Comparisons*

| BASE COMPARISON | PCS GF/GF | PCS HB/GF | SS exwGF/modGF | KEV GF/GF | KEV NIS/GF | NON HOMOGY | GF WIRES |
|---|---|---|---|---|---|---|---|
| LABORATORY | | | | | | | |
| BIO | | | | | | | |
| DHI | | | | | | | |
| IfM | | | | | | | |
| KORDI | | | | | | | |
| MRI | | | | | | | |
| NIES | | Sig GF>HB | Sig MOD>EXW | NS | Sig GF>NIS | | NS |
| NRL | | | | | | | |
| NISR (AA) | | | | | | | |
| NISR (ASV) | | | | | | | |
| SAG | | | | | | | |
| SKID | | | | | | | |
| TAMU | | | | | | | |
| UCON | | | | | | | |
| UDEL | | | | | | | |
| USM | | | | | | | |

## Table I.17a

*Zinc Numerical Comparison*
(μg l⁻¹)

| WIRE → / BOTTLE → | | PCS HB | PCS GF | SS mod GF | SS exw GF | KEV NIS | KEV GF | PCS ΣGF | SS ΣGF | KEV ΣGF |
|---|---|---|---|---|---|---|---|---|---|---|
| LABORATORY | | | | | | | | | | |
| BIO | m | 0.40 | 0.21 | 0.25 | 0.30 | 0.47 | 0.35 | 0.24 | 0.27 | 0.42 |
| | sd | 0.14 | 0.06 | 0.05 | 0.08 | 0.17 | 0.09 | 0.06 | 0.07 | 0.13 |
| DHI | m | | | | | | | | | |
| | sd | | | | | | | | | |
| IfM | m | | | | | | | | | |
| | sd | | | | | | | | | |
| KORDI | m | 3.2 | 3.3 | 7.1 | 3.3 | | | 3.3 | 5.2 | 3.8 |
| | sd | 2.1 | - | 0.5 | - | | | - | 2.3 | 0.8 |
| MRI | m | | | | | | | | | |
| | sd | | | | | | | | | |
| NIES | m | 1.24* | 0.062 | 0.124 | 0.303 | 0.316 | 0.183 | 0.062 | 0.222 | 0.260 |
| | sd | 0.74 | 0.003 | 0.010 | 0.050 | 0.115 | 0.037 | 0.003 | 0.098 | 0.112 |
| NRL | m | 0.47 | 0.70 | 0.30 | 0.20 | | | 0.70 | 0.25 | 0.18 |
| | sd | 0.46 | 0.22 | 0.14 | - | | | 0.22 | 0.10 | 0.18 |
| NISR (AA) | m | 1.53 | 1.2 | 0.55 | 1.4 | 1.2 | 0.75 | 1.2 | 0.91 | 0.61 |
| | sd | 0.39 | - | 0.17 | - | 0.4 | 0.57 | - | 0.47 | 0.41 |
| NISR (ASV) | m | | | | | | | | | |
| | sd | | | | | | | | | |
| SAG | m | 1.64 | 0.77 | | | 1.20 | 1.17 | 0.77 | | 1.14 |
| | sd | 1.18 | 0.08 | | | 0.30 | 0.05 | 0.08 | | 0.07 |
| SKID | m | 0.353 | 0.583 | 0.281 | 0.681 | 0.619 | 0.508 | 0.287 | 0.482 | 0.508 |
| | sd | 0.263 | 0.070 | 0.013 | 0.029 | 0.083 | 0.064 | 0.251 | 0.220 | 0.065 |
| TAMU | m | | | | | | | | | |
| | sd | | | | | | | | | |
| UCON | m | 0.543 | 0.329 | | | | | | | |
| | sd | 0.174 | 0.057 | | | | | | | |
| UDEL | m | | | | | | | | | |
| | sd | | | | | | | | | |
| USM | m | 4.6 | 4.3 | 5.9 | 7.9 | 10.5 | 5.7 | 4.3 | 6.7 | 5.7 |
| | sd | 0.6 | - | 1.1 | 1.3 | 1.0 | 1.2 | - | 1.5 | 1.1 |

## Table I.17b

*Zinc Statistical Comparisons*

| BASE COMPARISON | PCS GF/GF | PCS HB/GF | SS exwGF/modGF | KEV GF/GF | KEV NIS/GF | NON HOMOGY | GF WIRES |
|---|---|---|---|---|---|---|---|
| LABORATORY | | | | | | | |
| BIO | NS | Sig HB>GF | 90 EXW>MOD | Sig | NS | NS (2) | Sig KEV>SS>PCS |
| DHI | | | | | | | |
| IfM | | | | | | | |
| KORD | | NS | Sig MOD>EXW | | | | NS |
| MRI | | | | | | | |
| NIES | | Sig* HB>GF | Sig EXW>MOD | Sig | Sig NIS>GF | | Sig KEV>SS>PCS |
| NRL | | NS | NS | | | | Sig PCS>SS>KEV |
| NISR (AA) | | NS | Sig EXW>MOD | NS | Sig NIS>GF | | 90 PCS>SS>KEV |
| NISR (ASV) | | | | | | | |
| SAG | | NS | | NS | NS | | Sig KEV>PCS |
| SKID | Sig | NS | Sig EXW>MOD | NS | Sig NIS>GF | Sig (2) | 90 KEV>SS>PCS |
| TAMU | | | | | | | |
| UCON | Sig | NS | | | | | |
| UDEL | | | | | | | |
| USM | | NS | 90 EXW>MOD | NS | Sig NIS>GF | | Sig SS>KEV>PCS |

**Key to Tables I.8 - I.17**

| | |
|---|---|
| PCS | *Plastic-coated Steel* |
| SS | *Stainless Steel* |
| KEV | *Kevlar* |
| HB | *Hydro-Bios* |
| GF | *modified GO-FLO* |
| NIS | *Niskin* |
| modGF | *modified GO-FLO* |
| MOD | *modified GO-FLO* |
| exwGF | *ex-works GO-FLO* |
| EXW | *ex-works GO-FLO* |
| HOMOGY | *Homogeneity* |
| NS | *Not significant (P>0.1)* |
| 90 | *Significant (P<0.1)* |
| Sig | *Significant (P<0.05)* |
| m | *mean* |
| sd | *standard deviation* |
| ★ | *signifies suspected contamination* |
| (AA) | *Atomic Absorption* |
| (ASV) | *Anodic Stripping Voltametry* |
| BIO | *Bedford Institute of Oceanography* |
| DHI | *Deutsches Hydrographisches Institut* |
| IfM | *Institut fur Meereskunde* |
| KORDI | *Korea Ocean Research and Development Institute* |
| MRI | *Marine Research Institute* |
| NIES | *National Institute of Environmental Studies* |
| NISR | *Netherlands Institute for Sea Research* |
| NRL | *Naval Research Laboratory* |
| SAG | *Sagami Chemical Research Centre* |
| SKID | *Skidaway Institute of Oceanography* |
| TAMU | *Texas A&M University* |
| UCON | *University of Connecticut* |
| UDEL | *University of Delaware* |
| USM | *Universiti Sains Malaysia* |
| Sig (2) | *Significant water mass inhomogeneity detected during the second day's shipboard sampling operations.* |

## I.5 Ancillary studies

### Sampler Storage Experiments

Throughout the Bermuda Intercalibration Workshop, periodic meeting of the participants were held to discuss the day-to-day conduct of the programme. During these meetings we discussed additional studies that might be conducted to utilize the extra sample containers brought to Bermuda but not required since weather conditions forced a change to be made to the original plans for the core programme. The ancillary study unanimously agreed upon by the group was one to evaluate metal contamination of sea-water held for various lengths of time in different types of sampling bottle.

Sea water for this experiment was collected from about 2 metres below the surface about 2 km off the northeast coast of Bermuda using the research vessel PANULIRUS. Water was collected by pumping through 1/2'' diameter silicone rubber tubing using a peristaltic pump. Approximately 280 litres of sea water were pumped into a covered linear polyethylene vat of 600-litre capacity placed on the foredeck of the PANULIRUS.

After returning to the Bermuda Biological Station, water was pumped from the vat, using the same pumping system, into sample containers provided by the participants. Immediately afterwards, the three types of water sampler were also filled. Water samples were collected from each sampler after periods of a few minutes (early), 2 hours (mid), and 6 hours (late).

The participants in this experiment were DHI, IfM, UDEL, UCON, NRL, SKID and MRI (see key to Table I.4). Cadmium, copper, manganese, mercury, nickel and zinc results are given in Tables I.18 through I.23 and are briefly summarized below.

The initial 'tank' sea water used in this experiment should have lower cadmium concentrations than the deeper intercalibration samples since the surface waters are expected to be depleted in trace metals relative to deeper waters. UDEL and SKID results clearly show this (Table I.18). Results from these laboratories suggest that cadmium may increase with storage in GO-FLO bottles. The two sets of results for Hydro-Bio bottles are at variance. SKID results show no substantial increase in cadmium concentrations whereas DHI's results suggest a large initial increase in concentration.

Results for copper (Table I.19) indicate no significant increases with storage time in the GO-FLO sampler with the possible exception of a slight increase after six hours shown in the results of SKID. Both SKID and DHI report data that suggest that copper concentrations increase with time in the Hydro-Bios bottle.

One laboratory (DHI) examined changes in manganese concentrations in the Hydro-Bios bottle (Table I.20). These data show no substantial changes over the six-hour period of the storage experiment.

The data for mercury changes in the GO-FLO sampler obtained by IfM (Table I.21) are questionable and no conclusions can be drawn. The results from MRI suggest that the surface sample collected for this experiment has about the same mercury concentration as the deeper Panulirus Station water. MRI data also suggest that mercury concentrations increase with time in the Hydro-Bios bottle.

The results for nickel concentration changes (Table I.22) in the GO-FLO sampler are difficult to assess since the three laboratories reporting data do not agree. UDEL shows an increase in nickel with time whereas UCON reports a decrease. The results of SKID for the GO-FLO sampler are questionable but both SKID and DHI have provided results that reflect substantial nickel concentration increases with storage time in the Hydro-Bios bottle.

**Table I.18**

*Metal Change with Storage Time in Sampler. (CADMIUM.) (µg l⁻¹)*

| Participant | Sample | Time | GO-FLO | Time | HYDRO-BIOS | Time | NISKIN |
|---|---|---|---|---|---|---|---|
| DHI | Tank | | | | 0.009 | | |
| | Early | | | 6m | 0.048 ± 0.004 | | |
| | Mid | | | 1h 44m | 0.027 ± 0.0 | | |
| | Late | | | 6h 0m | 0.027 ± 0.004 | | |
| IfM | Tank | — | 0.019 ± 0.001 | | | | |
| | Early | 6m | 0.024 ± 0.013 | | | | |
| | Mid | 2h 1m | 0.020 ± 0.004 | | | | |
| | Late | 5h 58m | 0.019 ± 0.004 | | | | |
| NRL | Tank | | | | | — | — |
| | Early | | | | | 6m | 0.8 |
| | Mid | | | | | 2h 6m | 2.8 ± 0.4 |
| | Late | | | | | 6h 6m | 2.5 ± 1.1 |
| SKID | Tank | — | 0.0050 ± 0.0016 | — | — | | |
| | Early | 5m | 0.0058 ± 0.0027 | 6m | 0.011 — | | |
| | Mid | 2h 7m | 0.0094 ± 0.0023 | 1h 46m | 0.010 ± 0.0039 | | |
| | Late | 6h 34m | 0.0086 ± 0.0034 | 6h 0m | 0.0054 — | | |
| UCON | Tank | — | 0.033 ± 0.013 | | | | |
| | Early | 15m | 0.050 — | | | | |
| | Mid | 2h 25m | 0.021 ± 0.004 | | | | |
| | Late | 7h 9m | 0.031 ± 0.020 | | | | |
| UDEL | Tank | — | 0.0053 ± 0.0021 | | | | |
| | Early | 8m | 0.0044 ± 0.0006 | | | | |
| | Mid | 2h 16m | 0.0062 ± 0.0023 | | | | |
| | Late | 6h 42m | — | | | | |
| MRI | Tank | | | — | 0.011 ± 0.0 | — | 0.011 ± 0.0 |
| | Early | | | 3m | 0:015 ± 0.004 | 3m | 0.014 ± 0.001 |
| | Mid | | | 1h 47m | 0.019 ± 0.0 | 1h 59m | 0.016 ± 0.001 |
| | Late | | | 5h 56m | 0.014 ± — | 6h 10m | 0.014 ± 0.001 |

Excluding one apparently contaminated sample, (>l µg l⁻¹), the zinc results (Table I.23) reported by SKID suggest a significant increase after six hours' storage in GO-FLO bottles. The variability in the UCON results does not permit us to draw any conclusions from these data. The results for the Hydro-Bios bottle reported by SKID reflect some zinc contamination in this sampler.

The only laboratory to evaluate the effects of sea water storage in Niskin bottles on the levels of trace metals was NRL. Unfortunately, the results from this laboratory are too unreliable to be useful.

## Sampling and Analytical Methods used in Participating Laboratories

During the course of the Workshop all trace-metal participants were given a questionnaire requesting details of their current sea water sampling and analytical methods. Tables I.24 and I.25 show the information obtained through these questionnaires. The purpose of this information is to enable readers to determine the analytical methods used to analyse samples collected during the Workshop and to assess the comparability of each participant's normal sampling practices in the context of the results of the comparison of sampling devices conducted as part of the core programme of the Workshop.

## I.6 Conclusions

In general, the data returned by the participants in the trace-metal component at the Workshop are of good quality. Indeed, the relative agreement between experienced laboratories is better than previous sea-water trace-metal intercomparisons with the exception of the most recent intercalibrations for cadmium and mercury (Thibaud 1980 ; Olafsson 1980) conducted by ICES on behalf of the Joint Monitoring Group of the Oslo and Paris Commissions. The Bermuda exercise was, however, the first international intercalibration to assess directly the influences of commonly-used sampling devices and hydrowires on the determinations of a number of trace metals in open-ocean waters. The largest bodies of data accumulated during the exercise are for cadmium, copper, nickel and zinc, only a relatively few participants having returned data for iron, lead, manganese, mercury, molybdenum and vanadium. With the exclusion of outlying values, of which there were generally only a few for each element, the overall mean concentrations for all metals, except lead, are comparable with contemporary consensus values for metal concentrations in intermediate and deep-ocean waters. This is particularly remarkable since several of the participants have limited experience in the analysis of some of the metals, for which they have reported data in this experiment, in offshore seawater.

**Table I.19**

*Metal Change with Storage Time in Sampler (COPPER) (μg l⁻¹)*

| Participant | Sample | Time | GO-FLO | Time | HYDRO-BIOS | Time | NISKIN |
|---|---|---|---|---|---|---|---|
| DHI | Tank | | | — | 0.34 | | |
| | Early | | | 6m | 0.93 | | |
| | Mid | | | 1h 44m | 0.50 ± 0.01 | | |
| | Late | | | 6h 0m | 1.00 ± 0.25 | | |
| IfM | Tank | — | 0.310 ± 0.015 | | | | |
| | Early | 6m | 0.336 ± 0.056 | | | | |
| | Mid | 2h 1m | 0.300 ± 0.015 | | | | |
| | Late | 5h 29m | 0.300 ± 0.012 | | | | |
| NRL | Tank | | | | | — | — |
| | Early | | | | | 6m | 1.2 |
| | Mid | | | | | 2h 6m | 0.07 ± 0.03 |
| | Late | | | | | 6h 6m | 0.25 ± 0.07 |
| SKID | Tank | — | 0.240 ± 0.028 | — | 0.240 ± 0.028 | | |
| | Early | 5m | 0.260 ± 0.028 | 6m | 0.295 ± 007 | | |
| | Mid | 2h 7m | 0.285 ± 0.007 | 1h 46m | 0.325 ± 0.021 | | |
| | Late | 6h 34m | 0.325 ± 0.007 | 6h 0m | 0.400 — | | |
| UCON | Tank | — | 0.327 ± 0.139 | | | | |
| | Early | 15m | 0.357 ± 0.139 | | | | |
| | Mid | 2h 25m | 0.154 ± 0.029 | | | | |
| | Late | 7h 9m | 0.225 ± 0.016 | | | | |
| UDEL | Tank | — | 0.179 ± 0.046 | | | | |
| | Early | 8m | 0.178 ± 0.021d | | | | |
| | Mid | 2h 16m | 0.188 ± 0.030 | | | | |
| | Late | 6h 42m | — | | | | |
| MRI | Tank | | | — | 0.177 ± 0.001 | — | 0.117 ± 0.001 |
| | Early | | | 3m | 0.229 ± 0.006 | 3m | 0.186 ± 0.009 |
| | Mid | | | 1h 47m | 0.242 ± 0.001 | 1h 59m | 0.171 ± 0.0 |
| | Late | | | 5h 56m | 0.433 ± — | 6h 10m | 0.177 ± 0.019 |

Large differences in reported values for trace-metal constituents of open-ocean waters have, in the past, often been attributed to the effects of sampling devices and their methods of deployment. This experiment reveals, however, that differences resulting from the application of differing sampling procedures, assuming that reasonable precautions are taken, are relatively small for most of the commonly analysed metals. With a few exceptions, alternative sampling bottles and hydrowires in common use by experienced marine laboratories do not give rise to large discrepancies in the results of metal analysis of samples collected with them. For some elements there do appear to be significant differences in the results obtained with different sampling procedures. For copper and nickel, Hydro-Bios bottles appear to yield greater concentrations than do modified GO-FLO samplers. Similarly, modifications made to ex-works GO-FLO samplers, to improve their integrity, do result in marked reductions in the concentrations of copper, nickel and zinc measured in sea-water samples. For these same metals, and perhaps also for iron, Plastic-coated Steel would seem to be the cleanest hydrowire of the three types compared. Except in the case of zinc, Niskin bottles that have been suitably modified appear to be capable of collecting samples of only slightly inferior integrity to those collected with modified GO-FLO samplers. Some of these conclusions are supported by the results of the ancillary experiment in which changes in metal concentrations in sea-water samples retained within individual bottle types were measured.

The results of the entire exercise indicate that sufficient competence in trace-metal determinations in sea-water exists among experienced marine laboratories to justify making available their experience to other laboratories wishing to develop comparable expertise. The experiment has also shown that sampling methods currently being used by most developed marine laboratories are adequate for trace-metal sampling in the ocean at concentrations now believed to prevail in open-ocean waters. We would therefore recommend that the IOC, with the collaboration of WMO, UNEP and other concerned international organizations, as appropriate, now consider convening other regional workshops in which training and assistance in the fields of marine sampling, trace-metal analysis and trace-metal geochemistry be made available to personnel and institutions in developing countries. This might provide a most expedient mechanism for the provision of such assistance and more rapidly provide widespread capability for the assessment of trace-metal baselines and contamination in regional seas and continental shelf environments.

**Table I.20**

*Metal Change with Storage Time in Sampler (MANGANESE) ($\mu g\ l^{-1}$)*

| Participant | Sample | Time | GO-FLO | Time | HYDRO-BIOS | Time | NISKIN |
|---|---|---|---|---|---|---|---|
| DHI | Tank | | | — | 0.17 | | |
| | Early | | | 6m | 0.16 | | |
| | Mid | | | 1h 44m | 0.10 ± 0.03 | | |
| | Late | | | 6h 0m | 0.17 ± 0.04 | | |
| MRI | Tank | | | — | 0.024 ± 0.001 | — | 0.024 ± 0.001 |
| | Early | | | 3m | 0.023 ± 0.001 | 3m | 0.020 ± 0.001 |
| | Mid | | | 1h 47m | 0.026 ± 0.003 | 1h 59m | 0.025 ± 0.001 |
| | Late | | | 5h 56m | 0.024 ± 0.0 | 6h 10m | 0.024 ± 0.001 |
| | Tank | | | | | | |
| | Early | | | | | | |
| | Mid | | | | | | |
| | Late | | | | | | |
| | Tank | | | | | | |
| | Early | | | | | | |
| | Mid | | | | | | |
| | Late | | | | | | |
| | Tank | | | | | | |
| | Early | | | | | | |
| | Mid | | | | | | |
| | Late | | | | | | |
| | Tank | | | | | | |
| | Early | | | | | | |
| | Mid | | | | | | |
| | Late | | | | | | |

**Table I.21**

*Metal Change with Storage Time in Sampler (MERCURUY) ($\mu g\ l^{-1}$)*

| Participant | Sample | Time | GO-FLO | Time | HYDRO-BIOS | Time | NISKIN |
|---|---|---|---|---|---|---|---|
| IfM | Tank | — | 20.0 ± 0.4 | | | | |
| | Early | 6m | 22.1 ± 9.7 | | | | |
| | Mid | 2h 1m | 23.1 ± 9.9 | | | | |
| | Late | 5h 29m | 11.4 ± 2.8 | | | | |
| MRI | Tank | | | — | 4.0 ± 0.1 | | |
| | Early | | | 6m | 5.3 ± 0.2 | | |
| | Mid | | | 1h 47m | 7.1 ± 2.7 | | |
| | Late | | | 5h 36m | 14.7 — | | |
| | Tank | | | | | | |
| | Early | | | | | | |
| | Mid | | | | | | |
| | Late | | | | | | |
| | Tank | | | | | | |
| | Early | | | | | | |
| | Mid | | | | | | |
| | Late | | | | | | |
| | Tank | | | | | | |
| | Early | | | | | | |
| | Mid | | | | | | |
| | Late | | | | | | |
| | Tank | | | | | | |
| | Early | | | | | | |
| | Mid | | | | | | |
| | Late | | | | | | |

**Table I.22**

*Metal Change with Storage Time in Sampler (NICKEL) (µg $l^{-1}$)*

| Participant | Sample | Time | GO-FLO | Time | HYDRO-BIOS | Time | NISKIN |
|---|---|---|---|---|---|---|---|
| DHI | Tank | | | — | 0.09 | | |
| | Early | | | 6m | 0.38 | | |
| | Mid | | | 1h 46m | 1.45 ± 0.08 | | |
| | Late | | | 6h 0m | 0.84 ± 0.13 | | |
| NRL | Tank | | | | | — | — |
| | Early | | | | | 6m | 2.8 |
| | Mid | | | | | 2h 6m | 3.5 0.5 |
| | Late | | | | | 6h 6m | 3.4 ± 0.1 |
| SKID | Tank | — | 0.28 ± 0.0 | — | 0.28 ± 0.0 | — | 0.28 ± 0.0 |
| | Early | 5m | 0.26 ± 0.0 | 6m | 0.26 | | |
| | Mid | 2h 7m | 0.61 ± 0.12 | 1h 46m | 0.91 ± 0.13 | | |
| | Late | 6h 34m | 0.34 ± 0.0 | 6h 0m | 0.69 | | |
| UCON | Tank | — | 0.126 ± 0.008 | | | | |
| | Early | 15m | 0.164 ± 0.058 | | | | |
| | Mid | 2h 25m | 0.095 ± 0.002 | | | | |
| | Late | 7h 9m | 0.089 ± 0.013 | | | | |
| UDEL | Tank | — | 0.050 ± 0.007 | | | | |
| | Early | 8m | 0.041 ± 0.004 | | | | |
| | Mid | 2h 16m | 0.070 ± 0.010 | | | | |
| | Late | 6h 42m | — | | | | |
| MRI | Tank | | | — | 0.14 0.01 | — | 0.14 0.01 |
| | Early | | | 6m | 0.31 0.11 | 3m | 0.15 0.005 |
| | Mid | | | 1h 47m | 0.95 0.02 | 1h 59m | 0.15 0.05 |
| | Late | | | 5h 56m | 0.99 0.0 | 6h 10m | 0.99 0.0 |

**Table I.23**

*Metal Change with Storage Time in Sampler (ZINC) (µg $l^{-1}$)*

| Participant | Sample | Time | GO-FLO | Time | HYDRO-BIOS | Time | NISKIN |
|---|---|---|---|---|---|---|---|
| NRL | Tank | | | | | — | — |
| | Early | | | | | 6m | d.1. |
| | Mid | | | | | 2h 6m | 0,4 ± 0.2 |
| | Late | | | | | 6h 6m | 0.4 ± 0.2 |
| SKID | Tank | — | 0.23 ± 0.11 | — | 0.23 ± 0.11 | | |
| | Early | 5m | 0.22 ± 0.03 | 6m | 0.87 — | | |
| | Mid | 2h 7m | 0.20* ± — | 1h 46m | 0.66* | | |
| | Late | 6h 34m | 0.37 ± 0.07 | 6h 0m | 0.61 | | |
| UCON | Tank | — | 0.51 ± 0.15 | | | | |
| | Early | 15m | 0.61 ± — | | | | |
| | Mid | 2h 25m | 0.39 ± 0.05 | | | | |
| | Late | 7h 9m | 0.51 ± 0.05 | | | | |
| | Tank | | | | | | |
| | Early | | | | | | |
| | Mid | | | | | | |
| | Late | | | | | | |
| | Tank | | | | | | |
| | Early | | | | | | |
| | Mid | | | | | | |
| | Late | | | | | | |
| | Tank | | | | | | |
| | Early | | | | | | |
| | Mid | | | | | | |
| | Late | | | | | | |

* *excludes one result that is suspected to be biased by contamination.*

**Table I.24**

*Questionnaire Results*

| PARTICI-PANT | ELEMENTS | SAMPLE STORAGE | | | | | SAMPLE COLLECTION | | | | |
|---|---|---|---|---|---|---|---|---|---|---|---|
| | | Bottle type | Bottle size | Cleaning procedure | Pre-sampling condition | Sample Storage conditions temp | Hydrowire type | Hydro-weight type | Mes-senger type | Sampler type | Sampler modifi-cations |
| BIO | Cd.Cu.Zn. Ni Mn.Fe | CPE | 2000ml | HCl/$H_2O^{up}$/ $H_2O^{up}$+HCl up/ | Super-Q + 5ml $HCl^{up}$ | + 2.5ml/l $HCl^{up}$ A (Ultrex) 12H | 302 S.S. | Plastic-coated Pb | Weighted PVC | GO-FLO | Silicone 'O' rings. All-teflon stopcocks. Teflon-coat Internally. |
| DHI | Cd.Cu.Fe. Mn. Ni | Teflon | 500ml | O.I.N $HNO_3$/ $H_2O^{up}$/to low blank | Empty | + 0.5ml/120 % HCl A (silica distilled) | 4mm Zn gal-vanised steel coated with lmm PE | 25 kg Fe coated with teflon | Ni-coated brass (Comm.) | H-B | — |
| IfM | Hg.Zn.Cd. Pb. Cu | Silica | 500ml | Hot conc $HNO_3/H_2O^d$ at pH=l with HCl | Empty but wet | + 2ml/l $HNO_3$ A (Ultrex) | S.S. | Plastic box filled with sand. | Brass (Comm) | GO-FLO | — |
| KORDKI | Cd.Cu.Ni. Zn. | PE | 1000ml | $H_2O^c$/O.S./ $HCl^d$/APDC / $H_2O^{pd}$ | Empty | + 5ml/l $HNO_3$ A (Merck RG 2xdist) | Nylon (CZ) S.S. (Pel) | | (Comm) | Niskin<br>Van Dorn | —<br>— |
| MRI | Cd.Cu.Mn. Fe Ni.Pb.Zn | CPE | 1000ml | Det/ 8M $HNO_3$/ 5 X $H_2O^Q$ | Empty bagged | + 2.5ml/l 6M HCl A (Redist. Merck 317) | Polyam-ide coated galvan-ised steel | Iron | Brass (painted) | H-B | Silicone 'O' rings Pre-cleaned with det. |
| NIES | Mo.V.Ni.Cu. Cd.Zn.Fe | CPE | 1000ml | (Ref 1) | Dilute $HNO_3$ | + lml/l 65 % $HNO_3$ 4°C (Ultrex or Q.D.) | Iron ? | Iron ? | S.S. | Niskin | |
| NRL | Cd.Fe.Zn. Cu. Ni | CPE | 450ml | 2 X $HNO_3^{up}$/ 2 X $H_2O^{up}$/ 2 wks ex. SW/dit $HNO_3$ | Empty | + lml/l 12 N $HNO_3$ A (silica distilled) | 304 S.S. | Rosette | Rosette | Niskin | — |
| SAG | Hg Cd.Cu.Zn. Pb. | Glass<br>CPE<br>CPE<br>CPE | 500ml<br>1000ml<br>2000ml<br>5000ml<br>250ml | $HNO_3$<br>(HCl)<br>(HCl)<br>HCl | Empty<br>Empty | 10ml 18N $H_2SO_4$ A<br>++ 10ml/l 12N HCL A<br>++ 10ml/l 12N HCL A<br>++ 10ml/l 12N HCL A | Zn pl. Fe<br>Zn pl. Fe | —<br>— | Brass (Comm)<br>Brass (Comm) | Niskin<br>Niskin | --<br>— |
| SKID | Cd.Cu.Zn. Ni. | FEP Teflon | 250ml | Hot $HNO_3$/ $H_2O$ rinse/ 0.05 % $HNO_3$ | 0.05 % $HNO_3$ | + 400 1/1 $HNO_3^{up}$ A (N.B.S.) | Kevlar | — | Brass (Go-Devil) (Comm.) | Go-Flo | Silicone 'O' rings. All-Teflon stopcocks. Teflon-coat Internally |
| TAMU | Cu.Ni.Cd. | CPE | 1000ml | Hot HCl (Ref l) | Empty | + lml/l 16N $HNO_3$ A (Baker) | Any +10m Kevlar | 4'' diam PVC pipe weighted with Pb | Brass dipped in plastic coating | Niskin | Silicone surgical rubber spring. Drain tap modified. |
| UCON | Cd.Cu.Ni. Zn. | CPE | 1000ml | $HCl^C/HNO_3$ C/ hot 2N $HNO_3$/ hot O.ln HCl/ $H_2O^{up}$ rinse | Empty bagged | + 2ml/l $HNO_3^{up}$ A Redist. Baker reagent grade | By hand or by pumping from ≤4m depth using all-Teflon CPE system with in-line serial filtration. Surface only. | | | | |
| UDEL | Cu.Cd.Ni | CPE | 1000ml | 6N $HNO_3$/ 2N $HNO_3$/ $H_2O^{pd}$/ Super Q + acid | Empty bagged | + 2ml/l 12N $HNO_3$ R | Kevlar | Lead in PVC pipe | Brass (Comm) painted | GO-FLO | Silicone 'O' rings |

## TABLE I. 25

*Questionnaire Results*

| PARTICI-PANT | FILTRATION PROCEDURES | | | | | SHIPBOARD OPERATIONS | | | | LABORATORY ANALYTICAL PROCEDURES | | | |
|---|---|---|---|---|---|---|---|---|---|---|---|---|---|
| | Norm. Y/N | Filt. type | Pore size | Holder type | Tubing type | Cleaning proc. | Cleaning facilities | Shipboard steps | Analytical procedure | Clean facilities | No. of Analyses | Instruments | Other elements |
| B10 | Yes (Pr) | Nucl. | 0.4 | Milpr. SI (poly.) | | Super Q | No | + HCl | APDC/MIBK (Ni.Cu.Zn. Cd, Fe) Oxine/MIBR (Al. Mn) | LFH<br>Clean room | 2 | PE5000/50 0 | Al |
| DHI | No | Nucl. | 0.4 | Plexl-glass | Teflon | $H_2O$ (F) $HNO_3$(Oth-ers) | Yes | + HCl | APDC/MIBK | LFH | 3 | PE432<br>HGA500<br>ASI | Hg |
| IfM | No | Nucl. | 0.4 | Milpr. | Teflon | HCl (6N) | Yes | + $HNO_3$ | Au-amalg. (Hg) ASV (oth-ers) | LFH<br>Clean bench | 2 | MAS-50<br>PAR 174A | None |
| KORDI | Yes (V) | Nucl. or Milpr. | 0.4<br>0.45 | Glass | Tygon | O.S./Acid/ $H_2O^{pd}$ | No | +$HNO_3$ | Fe-APDC co-precip. | No | — | IL-251 | Co.Pb.Hg. Fe. Mn |
| MRI | No (Pel) Yes (CZ) (V) (direct) | Nucl. | 0.4 | Nucl. Polycarb. + silicone gaskets | None | 4N $HNO_3$(F) Det/$H_2O^0$(h) | No | Pre-concentr. occasional-ly | Chelex-100 | No | 2 | Varian AA-6<br>BC-6<br>ASD-53 | Al |
| NIES | Yes | N/A | N/A | N/A | N/A | N/A | No | +$HNO_3$ | APDC/DDDC-$CHCl_3$-$HNO_3$ | LFN<br>Clean room | 2 | JA ICP (MK 11) | ? |
| NISR | | | | | | | | | | | | | |
| HRL | Yes | N/A | N/A | N/A | N/A | N/A | Yes | + $HNO_3$ | APDC/MIBK-$HNO_3$/HCl | Clean bench | 2 | PE373<br>PE503 AAS<br>Spectras-pan III | None |
| SAG | No (V) | Milpr. | 0.45 | Acrylic | PVC | HCl/$H_2O^{pd}$ (F)<br>HCl/$H_2O^{pd}$ (H)<br>Det/$H_2O^{pd}$ (H) | No | + HCl | Au-amalg. (Hg) DDTC-MIBK-HCl (Oth-ers) | No | 2/3 | NJA AA-I (MK II) | As |
| SKID | Yes | Nucl. | 0.4 | Milpr. (teflon) | SI | 10 % HCl/$H_2O^Q$ | Yes | + $HNO_3$ | Dit.-Chlor.-$HNO_3$ | Clean room | 3 | PE<br>HGA2200<br>AS-1<br>PE 403 | As.Hg |
| TAMU | — | Nucl. | ? | Plastic | ? | Acid IN | Yes | + $HNO_3$ | Fe-APDC Co-precip. | Clean room<br>Clean hood | 2 | PE306<br>HGA-2100 | Hg |
| UCON | Yes | Glass-fibre<br>Nucl. | 0.8<br>0.2 | Milpr. | Teflon<br>CPE | $HCl^C$/$HNO_3^C$/ hot 2N $HNO_3$/ hot IN $HNO_3$/ $H_2O^Q$ | No | + $HNO_3$ | APDC/DDDC-Freon-$HNO_3$ | Cleanroom | 3 | PE5000/50 0 | Hg.As. Cr. Sb |
| UDEL | Yes (Pr) Direc | Nucl. | 0.4 | Acrylic | CPE | 6N $HNO_3^h$/ 2N $HNO_3^h$/ Super-Q acidified | Yes | + $HNO_3$ | APDC/DDDC-Freon/$HNO_3$ | LFH | 2/3 | IL751 | Ba, Cu, Co, Cd, Ni, Fe, Mn, Mo |


## Acknowledgements

On behalf of the IOC, WMO and UNEP we should like to express our thanks to all those individuals and institutions that provided oceanographic and other equipment necessary for the conduct of the core and ancillary experiments. Especial thanks are due to those participants who braved the elements and so conscientiously carried out the shipboard sampling work. This latter group comprised Don Atwood, Lutz Brugmann, Cameron McLeod, Jon Olafsson, Steve Piotrowicz, Deither Schmidt, Ralph Smith, Frank Storti and Phil Yeats. Finally, we gratefully acknowledge the assistance of John L. Barron, who carried out the computer analyses of the data, and Judy Simms for so willingly typing and editing this report.

We particularly wish to thank the participants, without whom this experiment could not have been conducted. They are to be commended for their willingness to provide analytical data for blind samples in an intercalibration in which each laboratory is identified with its results. It should be stressed that several of the participants had limited experience, either in the analysis of certains metals or in sea-water analyses generally, and for them this exercise constitutes a learning experience. In no case should the results of any laboratory be taken to reflect its future capability to perform trace metal determinations in sea-water. Their participation in this experiment clearly indicates their dedication to the improvement of sampling and analytical capabilities in this field.

### *Key for Tables I. 24 and I. 25*

| | |
|---|---|
| CPE | *Conventional polyethylene* |
| PE | *Polyethylene (Unspecified type)* |
| $HC1^c$ | *Concentrated hydrochloric acid* |
| $H_2O^d$ | *Distilled water* |
| $H_2O^{up}$ | *Ultra-pure water* |
| $H_2O^{pd}$ | *Distilled de-ionised water* |
| $H_2O^Q$ | *Super Q water (Millipore)* |
| Q.D. | *Quartz distilled* |
| O.S. | *Organic solvent* |
| SW | *Sea water* |
| - | *Storage acid added prior to sampling* |
| + | *Storage acid added immediately after sampling* |
| ++ | *Storage acid added after samples returned to laboratory* |
| A | *Ambient temperature* |
| R | *Refrigerated* |
| pl. | *Plated* |
| S.S. | *Stainless steel* |
| 302 S.S. | *Type 302 stainless steel* |
| 304 S.S. | *Type 304 stainless steel* |
| CZ | *Coastal zone samples* |
| Pel | *Pelagic (deep ocean) samples* |
| H-B | *Hydro-Bios samplers* |
| (V) | *Vacuum filtration* |
| (Pr) | *Pressure filtration* |
| Milpr. | *Millipore* |
| Nucl. | *Nuclepore* |
| Milpr. Si | *Silicone 'O' ringed Millipore holder* |
| (poly) | *Polypropylene construction* |
| (teflon) | *Teflon construction* |
| Dith. | *Dithizone* |
| Chlor. | *Chloroform* |
| LFH | *Laminar flow hood* |
| (f) | *Filter* |
| (h) | *Holder* |
| NJA | *Nippon Jarrell Ash Co.* |
| PAR | *Princeton Applied Research Inc.* |
| IL | *Instrumentation Laboratory* |

## List of participants


| | |
|---|---|
| AMBE, M. | *Sagami Chemical Research Centre, Nish-Ohnuma 4-4-1, Sagamihara, Kanawaga 229, Japan* |
| ATWOOD, D. | *Atlantic Oceanographic and Meteorological Laboratory, National Oceanographic and Atmospheric Administration, 15, Rickenbacker Causeway, Miami, Florida 33149, USA* |
| BEWERS, J.M. | *Atlantic Oceanographic Laboratory, Bedford Institute of Oceanography, P.O. 1006, Dartmouth, Nova Scotia, Canada* |
| BRUGMANN, L. | *Institut fur Meereskunde, Akademie der Wissenschaften der DDR, Warnemunde, Rostock 253, German Democratic Republic* |
| CHURCH, T. | *College of Marine Studies, University of Delaware, Newark, Delaware 19711, USA* |
| DUINKER, J.C. | *Netherlands Institute of Sea Research, P.O. 59, Den Burg, Texel, Netherlands* |
| LEE, K. | *Korea Ocean Research and Development Institute, P.O. 131, Dongdaemun, Seoul, Korea* |
| McLEOD, C. | *National Institute for Environmental Studies, Yatabe, Ibaraki, Japan* |
| OLAFSSON, J. | *Marine Research Institute, Skulagata 4, Reykjavik, Iceland* |
| PELLENBARG, R. | *Naval Research Laboratory, Code 4330, Washington, D.C. 20375, USA* |
| PIOTROWICZ, S. | *Atlantic Oceanographic and Meteorological Laboratory, National Oceanographic and Atmospheric Administration, 15, Rickenbacker Causeway, Miami, Florida 33149, USA* |
| PRESLEY, B. | *Department of Oceanography, Texas A&M University, College Station, Texas 77843, USA* |
| SCHMIDT, D. | *Deutsches Hydrographisches Institut, Laboratorium Sulldorf, Df20 Hamburg 55, Wustland 2, Federal Republic of Germany* |
| SIVALINGAM, M. | *School of Biological Sciences, University of Sains Malaysia, Penang, Malaysia* |
| SMITH, R. | *Skidaway Institute of Oceanography, Savannah, Georgia 31406, USA* |
| STORTI, F. | *Skidaway Institute of Oceanography, Savannah, Georgia 31406, USA* |
| TRAMONTANO. J. | *College of Marine Studies, University of Delaware, Newark, Delaware 19711, USA* |
| WASLENCHUK, D. | *Marine Sciences Institute, University of Connecticut, Avery Point, Groton, Connecticut 06340, USA* |
| WINDOM, H.L. | *Skidaway Institute of Oceanography, P.O. 13687, Savannah, Georgia 31406, USA* |
| YEATS, P.A. | *Atlantic Oceanographic Laboratory, Bedford Institute of Oceanography, P.O. 1006, Dartmouth, Nova Scotia, Canada* |

## References


*BEWERS, J.M., G. TOPPING and H.L. WINDOM. 1978. « Status and plans regarding ICES intercalibrations for trace metals in sea water », ICES CM 1978/E :27.*

*BEWERS, J.M., J. DALZIEL, P.A. YEATS and J.L. BARRON. 1979. « Report of the ICES Fourth Round Intercalibration for Trace Metals in Sea water », ICES CM1979/E :37.*

*BEWERS, J.M., P.G.W. JONES, K. KREMLING, J. OLAFSSON, G. TOPPING and H.L. WINDOM. 1980. « Design and logistical options for the conduct of the ICES Fifth Round Intercalibration for Trace Metals in Sea Water », ICES CM1980/E :19.*

*ICES. 1980. « Second report of the Marine Chemistry Working Group », Copenhagen, 12-14 Feb. 1980. CM1980/C :1.*

*IOC. 1976. « A comprehensive plan for the global investigation of pollution of the marine environment and baseline study guidelines », IOC Technical Series 14, Unesco Paris.*

*IOC/UNEP/WMO. 1976. « A programme for monitoring background levels of selected pollutants in open-ocean waters », Unesco, Paris.*

*JONES, P.G.W. 1976. « An ICES intercalibration exercise for trace metal standard solutions ». ICES CM1979/E :15.*

*JONES, P.G.W. 1977. « A preliminary report on the ICES intercalibration of sea water samples for the analyses of trace metals ». ICES CM1977/E :16.*

*LANDING, W.M. and K.W. BRULAND. 1980. « Manganese in the North Pacific »,* Earth and Planet. Sci. Lett., 49 *: 45-56.*

*NIE, N.H., C.H. HULL, J.G. JENKINS, K. STEINBRENNER and D.H. BENT. 1975. « Statistical package for the social sciences ». Second Edition. McGraw-Hill, Inc., New York.*

*OLAFSSON, J. 1976. « Report on ICES international intercalibration on mercury in sea water », ICES CM1976/E :49.*

*OLAFSSON, J. 1980. « A preliminary report on ICES intercalibration of mercury in sea water for the Joint Monitoring Group of the Oslo and Paris Commissions », Submitted to the Marine Chemistry Working Groupe of ICES, Feb. 1980.*

*POCKLINGTON, R. 1972a. « Variability in the ocean off Bermuda », Bedford Institute of Oceanography Report Series/B1-R-72-3/May 1972.*

*POCKLINGTON, R. 1972b. « Secular changes in the ocean off Bermuda »,* J. Geophys. Res., 77 *: 6604-6607.*

*THIBAUD, Y. 1980. « Exercice d'intercalibration CIEM 1979, cadmium en eau de mer », Report submitted to the Marine Chemistry Working Group of ICES, Feb. 1980.*

*WINDOM, H.L. 1979. « Report on the results of the ICES questionnaire on sampling and analysis of sea water for trace elements », Submitted to the First Meeting of the Marine Chemistry Working Group, Lisbon, May 1979.*

# Results received from participating laboratories

## BEDFORD INSTITUTE OF OCEANOGRAPHY

ADDRESS : Bedford Institute of Oceanography
P.O. Box 1006
Dartmouth, Nova Scotia B2Y 4A2
Canada

ACRONYM : BIO

### CADMIUM
($\mu g\ l^{-1}$)

| | | | | |
|---|---|---|---|---|
| 0530 | 0.046 | 0.036 | | |
| | 0.038 | 0.055 | | |
| | 0.049 | 0.051 | 0.061 | 0.063 |
| | 0.049 | 0.042 | | |
| 0531 | 0.051 | 0.040 | | |
| | 0.075 | 0.104 | 0.128 | |
| | 0.044 | 0.056 | 0.099 | 0.105 |
| 0532 | 0.051 | 0.04 | | |
| | 0.072 | | | |
| | 0.030 | 0.055 | | |
| | 0.060 | 0.067 | 0.040 | |
| 0549 | 0.045 | | | |
| | 0.061 | 0.062 | | |
| | 0.059 | 0.044 | | |
| | 0.042 | 0.045 | | |
| 0551 | 0.032 | 0.029 | | |
| | 0.033 | 0.041 | | |
| | 0.056 | 0.043 | | |
| 0570 | 0.041 | 0.033 | | |
| | 0.044 | 0.036 | | |
| | 0.043 | 0.042 | | |
| 0571 | 0.060 | 0.050 | 0.043 | 0.038 |
| 0572 | 0.043 | 0.058 | 0.062 | |
| 0573 | 0.036 | 0.032 | | |
| 0574 | 0.042 | | | |
| 0579 | 0.048 | 0.036 | | |
| | 0.047 | 0.047 | | |
| | 0.046 | 0.039 | | |
| 0613 | 0.039 | 0.034 | | |
| | 0.060 | 0.050 | | |
| 0621 | 0.081 | 0.016 | | |
| | 0.055 | 0.014 | | |
| | 0.029 | 0.038 | | |

### COPPER
($\mu g\ l^{-1}$)

| | | | | |
|---|---|---|---|---|
| 0530 | 0.114 | 0.096 | | |
| | 0.114 | 0.125 | | |
| | 0.110 | 0.100 | | |
| | 0.103 | 0.126 | | |
| 0531 | 0.115 | 0.123 | | |
| | 0.129 | 0.120 | 0.138 | 0.130 |
| | 0.138 | 0.158 | 0.125 | 0.137 |
| 0532 | 0.113 | 0.100 | | |
| | 0.105 | 0.147 | | |
| | 0.068 | 0.088 | | |
| | 0.115 | 0.114 | | |

### COPPER (continued)

| | | | | |
|---|---|---|---|---|
| 0549 | 0.099 | 0.102 | | |
| | 0.082 | 0.100 | | |
| | 0.112 | 0.098 | | |
| | 0.076 | 0.088 | | |
| 0551 | 0.100 | 0.093 | | |
| | 0.130 | 0.101 | | |
| | 0.095 | 0.105 | | |
| | 0.104 | 0.092 | | |
| 0570 | 0.107 | 0.086 | | |
| | 0.096 | 0.082 | | |
| | 0.084 | 0.094 | | |
| 0571 | 0.095 | 0.086 | | |
| 0572 | 0.312 | 0.326 | 0.430 | 0.430 |
| 0573 | 0.088 | 0.091 | | |
| 0574 | 0.106 | 0.095 | | |
| 0579 | 0.102 | 0.076 | | |
| | 0.117 | 0.104 | | |
| | 0.079 | 0.095 | | |
| 0613 | 0.098 | 0.094 | | |
| | 0.109 | 0.096 | 0.094 | 0.102 |
| 0621 | 0.090 | 0.081 | | |
| | 0.102 | 0.070 | | |
| | 0.084 | 0.092 | | |

### IRON
($\mu g\ l^{-1}$)

| | | | | |
|---|---|---|---|---|
| 0530 | 0.42 | 0.32 | | |
| | 0.32 | 0.30 | | |
| | 0.70 | 0.65 | 0.64 | 0.52 |
| | 0.69 | 0.72 | | |
| 0531 | 0.29 | 0.28 | | |
| | 0.49 | 0.55 | 0.51 | 0.31 |
| | 0.31 | 0.34 | 0.38 | 0.39 |
| 0532 | 0.56 | 0.49 | | |
| | 0.51 | 0.55 | | |
| | 0.62 | 0.62 | | |
| | 0.74 | 0.71 | 0.82 | 0.82 |
| 0549 | 0.51 | 0.55 | | |
| | 1.71 | 1.68 | 1.89 | 2.04 |
| | 1.89 | 1.99 | 1.86 | 2.04 |
| | 0.42 | 0.41 | 0.31 | 0.34 |
| 0551 | 0.48 | 0.47 | | |
| | 0.52 | 0.53 | | |
| | 0.48 | 0.53 | 0.31 | 0.36 |
| | 0.37 | 0.29 | | |

## IRON (continued)

| | | | | |
|---|---|---|---|---|
| 0570 | 1.00 | 0.96 | | |
| | 0.86 | 0.84 | | |
| | 0.43 | 0.52 | | |
| 0571 | 0.39 | 0.34 | 0.40 | 0.42 |
| 0572 | 2.60 | 2.40 | 3.34 | 3.47 |
| 0573 | 0.26 | 0.28 | | |
| 0574 | 0.84 | 0.68 | | |
| 0579 | 0.68 | 0.51 | | |
| | 0.51 | 0.42 | | |
| | 0.47 | 0.52 | | |
| 0613 | 0.47 | 0.48 | | |
| | 0.39 | 0.34 | | |
| 0621 | 0.38 | 0.37 | | |
| | 0.44 | 0.35 | | |
| | 0.24 | 0.24 | | |

## NICKEL
**(μg l$^{-1}$)**

| | | | | |
|---|---|---|---|---|
| 0530 | 0.257 | 0.224 | | |
| | 0.232 | 0.249 | | |
| | 0.260 | 0.192 | | |
| | 0.216 | 0.230 | | |
| 0531 | 0.198 | 0.205 | | |
| | 0.197 | 0.169 | 0.239 | 0.259 |
| | 0.187 | 0.177 | 0.233 | 0.239 |
| 0532 | 0.214 | 0.220 | | |
| | 0.223 | 0.223 | | |
| | 0.184 | 0.180 | | |
| | 0.191 | 0.187 | | |
| 0549 | 0.236 | 0.245 | | |
| | 0.200 | 0.215 | | |
| | 0.186 | 0.181 | | |
| | 0.200 | 0.211 | | |
| 0551 | 0.253 | 0.245 | | |
| | 0.250 | 0.248 | | |
| | 0.255 | 0.215 | | |
| | 0.232 | 0.242 | | |
| 0570 | 0.194 | 0.181 | | |
| | 0.230 | 0.211 | | |
| | 0.219 | 0.208 | | |
| 0571 | 0.219 | 0.216 | 0.204 | 0.223 |
| 0572 | 0.405 | 0.400 | 0.395 | 0.429 |
| 0573 | 0.232 | 0.220 | | |
| 0574 | 0.252 | | | |
| 0579 | 0.225 | 0.190 | | |
| | 0.223 | 0.224 | | |
| | 0.185 | 0.172 | | |
| 0613 | 0.245 | 0.251 | | |
| | 0.228 | 0.230 | 0.219 | 0.216 |
| 0621 | 0.161 | 0.165 | | |
| | 0.167 | 0.157 | | |
| | 0.161 | 0.167 | | |

## MANGANESE
**(μg l$^{-1}$)**

| | | | | |
|---|---|---|---|---|
| 0530 | 0.049 | 0.054 | | |
| | 0.051 | 0.043 | | |
| | 0.053 | 0.052 | | |
| | 0.049 | 0.072 | | |
| 0531 | 0.053 | 0.051 | | |
| | 0.048 | 0.055 | | |
| | 0.051 | 0.056 | | |

## MANGANESE (continued)

| | | | | |
|---|---|---|---|---|
| 0532 | 0.045 | 0.048 | | |
| | 0.050 | 0.046 | | |
| | 0.061 | 0.057 | | |
| 0549 | 0.028 | 0.028 | | |
| | 0.042 | 0.045 | | |
| | 0.050 | | | |
| | 0.067 | 0.067 | | |
| 0551 | 0.041 | 0.031 | | |
| | 0.033 | 0.032 | | |
| | 0.035 | 0.041 | | |
| | 0.059 | 0.060 | | |
| 0570 | 0.038 | 0.038 | | |
| | 0.130 | 0.139 | | |
| | 0.090 | 0.113 | | |
| 0571 | 0.047 | 0.039 | | |
| 0572 | 0.435 | 0.765 | | |
| 0573 | 0.060 | 0.063 | | |
| 0574 | 0.028 | | | |
| 0579 | 0.025 | 0.023 | | |
| | 0.078 | 0.098 | | |
| | 0.049 | 0.060 | | |
| 0613 | 0.036 | 0.030 | | |
| | 0.050 | 0.048 | | |
| 0621 | 0.047 | 0.055 | | |
| | 0.046 | 0.049 | | |
| | 0.064 | 0.063 | | |

## ZINC
**(μg l$^{-1}$)**

| | | | | |
|---|---|---|---|---|
| 0530 | 0.67 | 0.28 | | |
| | 0.40 | 0.50 | | |
| | 0.52 | 0.54 | 0.62 | 0.69 |
| | 0.32 | 0.18 | | |
| 0531 | 0.31 | 0.27 | | |
| | 0.28 | 0.23 | | |
| | 0.36 | 0.41 | 0.48 | 0.42 |
| 0532 | 0.58 | 0.46 | | |
| | 0.48 | 0.61 | | |
| | 0.26 | 0.55 | | |
| | 0.27 | 0.38 | 0.58 | 0.54 |
| 0549 | 0.22 | 0.33 | 0.24 | |
| | 0.26 | 0.18 | 0.29 | 0.31 |
| | 0.22 | 0.25 | 0.17 | 0.20 |
| | 0.25 | 0.28 | 0.30 | |
| 0551 | 0.51 | 0.30 | | |
| | 0.25 | 0.28 | 0.21 | |
| | 0.33 | 0.22 | 0.24 | 0.34 |
| | 0.32 | 0.27 | | |
| 0570 | 0.22 | 0.22 | | |
| | 0.24 | 0.30 | | |
| | 0.14 | 0.14 | | |
| 0571 | 0.26 | 0.22 | | |
| 0572 | 1.55 | 1.58 | 1.10 | 1.30 |
| 0573 | 0.51 | 0.52 | | |
| 0574 | 0.54 | 0.37 | | |
| 0579 | 0.28 | 0.18 | | |
| | 0.28 | 0.21 | | |
| | 0.35 | 0.14 | | |
| 0613 | 0.31 | 0.29 | | |
| | 0.26 | 0.22 | 0.19 | 0.31 |
| 0621 | 0.27 | 0.21 | | |
| | 0.31 | 0.17 | | |
| | 0.36 | 0.24 | | |

| | Nickel | Iron | Zinc | Cadmium | Copper | Manganese | Aluminium |
|---|---|---|---|---|---|---|---|
| 0530 | .233 ± .023 | .53 ± .17 | .47 ± .17 | .049 ± .009 | .111 ± .011 | .052 ± .009 | 1.10 ± .48 |
| 0531 | .210 ± .030 | .39 ± .10 | .37 ± .11 | .053 ± .014* | .131 ± .012 | .052 ± .003 | 1.11 ± .15 |
| 0532 | .203 ± .019 | .64 ± .12 | .47 ± .13 | .052 ± .014 | .106 ± .023 | .050 ± .006 | 2.21 ± .58 |
| 0549 | .209 ± .022 | .42 ± .09* | .27 ± .10 | .051 ± .009 | .095 ± .012 | .047 ± .016 | 1.01 ± .13* |
| 0551 | .243 ± .013 | .43 ± .09 | .29 ± .08 | .039 ± .010 | .102 ± .013 | .042 ± .012 | 1.09 ± .19 |
| 0570 | .207 ± .018 | .77 ± .24 | .21 ± .06 | 0.040 ± .004 | .092 ± .009 | .091 ± .045 | 1.38 ± .34 |
| 0571-4 | .224 ± .015* | .45 ± .20* | .40 ± .14* | .046 ± .011 | .094 ± .007* | .047 ± .015* | .91 ± .17* |
| 0579 | .203 ± .024 | .52 ± .09 | .24 ± .08 | .044 ± .005 | .096 ± .016 | .056 ± .030 | 1.13 ± .16 |
| 0613 | .232 ± .014 | .42 ± .07 | .26 ± .05 | .046 ± .012 | .099 ± .006 | .041 ± .010 | 1.47 ± .17 |
| 0621 | .163 ± .004 | .34 ± .08 | .24 ± .09 | .039 ± .026 | .087 ± .011 | .054 ± .008 | 1.08 ± .11 |
| 0624 | .185 ± .013 | 1.65 ± .02 | .46 ± .21 | .037 ± .016 | .817 ± .047 | | |

*Units $\mu g\ l^{-1}$

## DEUTSCHES HYDROGRAPHISCHES INSTITUT

ADDRESS: Deutsches Hydrographisches Institut
Postfach 2 20
2000 Hamburg 4
Federal Republic of Germany

ACRONYM: DHI

| Sampler code no. | Our bottle no. | Cd | Cu | Fe | Mn | Ni |
|---|---|---|---|---|---|---|
| 800 630 | 328 | 0.027 | 0.51 | 2.78 | 0.08 | 1.39 |
| | | | | | 0.12 | |
| | 357 | 0.027 | 0.49 | 15.8 | 0.15 | 1.50 |
| | | | | | (0.23) | |
| | | | | | (0.25) | |
| 800 632 | 330 | 0.024 | 0.83 | 8.10 | 0.14 | 0.75 |
| | 331 | 0.030 | 1.18 | 10.3 | 0.19 | 0.93 |
| 800 627 | 319 | 0.050 | 0.93 | 6.47 | 0.16 | 0.38 |
| | | 0.045 | | | | |
| | 360 | 0.019 | 0.38 | 7.86 | 0.15 | 0.31 |
| 800 623 | 355 | 0.009 | 0.34 | 15.5 | (0.06) | 0.09 |
| | | | | | 0.17 | |
| | 356 | 0.011 | 0.50 | 248.0 | 0.62 | 0.14 |
| 800 601 | 300 | 0.028 | 0.67 | 28.6 | 0.32 | 0.21 |
| | 333 | 0.025 | 0.70 | 16.3 | 0.11 | |
| | | | | | ≪ 0.12 | |
| | 301 | 0.008 | 0.51 | 7.81 | 0.24 | 0.14 |
| | 324 | 0.008 | 1.18 | 34.6 | 0.20 | 0.36 |
| | | | | | 0.21 | |
| 800 545 | 308 | 0.075 | 0.53 | 12.7 | 0.05 | 0.26 |
| | 323 | 0.058 | 0.62 | 6.91 | 0.09 | 0.29 |
| | | | | | (0.04) | |
| | 329 | 0.101 | 0.57 | 12.3 | 0.06 | 0.20 |
| | 332 | 0.094 | 0.76 | 18.8 | 0.20 | 0.19 |
| 800 543 | 303 | 0.099 | 0.30 | 4.43 | 0.03 | 0.20 |
| | 305 | 0.158 | 0.36 | 17.7 | 0.08 | 0.20 |
| | 3 | 0.075 | 0.82 | 15.5 | 0.07 | 0.35 |
| | 327 | 0.041 | 0.73 | 2.71 | < 0.05 | 0.19 |
| 800 525 | 318 | 0.024 | 1.18 | 8.81 | 0.08 | 0.23 |
| | 336 | 0.034 | 0.30 | 9.55 | 0.06 | 0.22 |
| | 352 | 0.040 | 0.17 | 2.94 | 0.03 | 0.23 |
| | n.N. | 0.048 | 0.17 | 3.23 | 0.03 | 0.19 |
| 800 526 | 304 | 0.028 | 0.19 | 3.19 | 0.06 | 0.16 |
| | | | | | (0.04) | |
| | 326 | 0.029 | 0.57 | 9.48 | 0.04 | 0.13 |
| | 353 | 0.041 | 0.13 | 4.02 | 0.06 | 0.17 |
| | | 0.049 | | | | |
| | 359 | 0.031 | 0.14 | 13.3 | 0.03 | 0.14 |
| | | | 0.33 | | | |

| Sampler code no. | Our bottle no. | Cd | Cu | Fe | Mn | Ni |
|---|---|---|---|---|---|---|
| 800 527 | 302 | 0.034 | 0.26 | 10.8 | 0.02 | 0.15 |
| | 320 | 0.023 | 0.73 | 3.06 | 0.01 | 0.42 |
| | | | (1.11) | | | |
| | 325 | 0.026 | 0.43 | 4.54 | 0.07 | 0.29 |
| | 351 | 0.049 | 0.28 | 6.32 | 0.10 | 0.39 |
| | | | (0.26) | | | |
| 800 597 | 335 | 0.051 | 0.56 | 2.19 | 0.02 | 0.28 |
| 800 598 | 354 | 0.035 | 2.16 | 2.90 | 0.04 | 0.30 |
| 800 600 | 322 | 0.014 | 0.19 | 3.03 | 0.06 | 0.27 |
| 800 597<br>800 598<br>800 600 | 334 | 0.031 | 1.09 | 3.13 | < 0.04 | 0.34 |

*Units µg l⁻¹*

# INSTITUT FUR MEERESKUNDE

ADDRESS : Institut fur Meereskunde
Akademie der Wissenschaften der DDR
Warnemunde, Rostock 253
German Democratic Republic

ACRONYM : IfM

| Sample No. | Hg | | Cd | | Cu | |
|---|---|---|---|---|---|---|
| 530/1 | 34.7 ; | 36.0 | 996 ; | 973 | 1333 ; | 1290 |
| 530/2 | 5.5 ; | 6.2 | 41 ; | 37 | 86 ; | 77 |
| 530/3 | 6.5 ; | 6.3 | 643 ; | 609 | 293 ; | 290 |
| 530/4 | 3.2 ; | 3.8 | 284 ; | 302 | 109 ; | 99 |
| 531/1 | 5.3 ; | 5.5 | 778 ; | 750 | 819 ; | 830 |
| 531/2 | 5.6 ; | 5.0 | 51 ; | 50 | 281 ; | 270 |
| 532/1 | 6.8 ; | 7.2 | 1573 ; | 1490 | 572 ; | 562 |
| 532/2 | 16.9 ; | 17.5 | 513 ; | 527 | 816 ; | 832 |
| 532/3 | 3.8 ; | 4.1 | 58 ; | 62 | 121 ; | 116 |
| 532/4 | 9.4 ; | 10.0 | 53 ; | 58 | 115 ; | 122 |
| 556/1 | 3.6 ; | 3.8 | 108 ; | 121 | 159 ; | 175 |
| 556/2 | 22.7 ; | 23.6 | 34 ; | 40 | 208 ; | 187 |
| 556/3 | 3.9 ; | 3.6 | 46 ; | 40 | 156 ; | 170 |
| 556/4 | 3.2 ; | 3.5 | 60 ; | 67 | 309 ; | 322 |
| 558/1 | 19.5 ; | 20.2 | 4770 ; | 4020 | 261 ; | 251 |
| 558/2 | 1.5 ; | 1.6 | 25 ; | 27 | 176 ; | 182 |
| 558/3 | 1.5 ; | 1.5 | 19 ; | 23 | 213 ; | 200 |
| 558/556 | 9.7 ; | 10.2 | 50 ; | 52 | 176 ; | 178 |
| 575/1 | 2.6 ; | 2.7 | 24 ; | 28 | 167 ; | 162 |

| Sample No. | Hg | | Cd | | Cu | |
|---|---|---|---|---|---|---|
| 575/2 | 2.6 ; | 2.5 | 36 ; | 31 | 122 ; | 125 |
| 576/1 | 4.1 ; | 4.3 | 49 ; | 52 | 463 ; | 493 |
| 576/2 | 18.0 ; | 17.6 | 224 ; | 240 | 1008 ; | 988 |
| 577 | 2.1 ; | 2.2 | 53 ; | 57 | 739 ; | 709 |
| 578 | 7.2 ; | 7.5 | 25 ; | 22 | 135 ; | 137 |
| 579/1 | 7.5 ; | 7.4 | 36 ; | 33 | 156 ; | 150 |
| 579/2 | 22.7 ; | 23.2 | 138 ; | 133 | 285 ; | 282 |
| 579/3 (PTFE) | 0.7 ; | 1.1 | 34 ; | 35 | 115 ; | 89 |
| 623/1 (PTFE) | 1.7 ; | 2.0 | 10 ; | 13 | 260 ; | 262 |
| 623/2 | 3.4 ; | 3.6 | 23 ; | 20 | 299 ; | 280 |
| 623/624 | 20.3 ; | 19.7 | 18 ; | 20 | 321 ; | 300 |
| 625/1 | 13.3 ; | 14.1 | 11 ; | 15 | 283 ; | 294 |
| 625/2 | 30.1 ; | 31.0 | 34 ; | 36 | 376 ; | 392 |
| 628/1 | 31.3 ; | 32.0 | 22 ; | 24 | 293 ; | 286 |
| 628/2 | 14.7 ; | 14.2 | 18 ; | 15 | 320 ; | 300 |
| 633/1 | 13.5 ; | 14.1 | 13 ; | 17 | 311 ; | 309 |
| 633/2 | 8.7 ; | 9.2 | 23 ; | 21 | 286 ; | 295 |

*Units : ng l⁻¹, pH ≈ 2*

# KOREA OCEAN RESEARCH & DEV. INSTITUTE

ADDRESS : Korea Ocean Research & Dev. Institute
P.O. Box 131
Dongdaemun, Seoul
Korea

ACRONYM : KORDI

| Sample No. | Zn | Cu | Cd | Ni |
|---|---|---|---|---|
| 800 504 | 6.87 | 1.13 | 0.04 | 0.23 |
| 504 | 7.75 | 1.50 | 0.07 | 0.47 |
| 505 | 4.38 | 0.94 | 0.04 | 0.23 |
| 505 | 3.25 | 1.19 | 0.07 | 0.23 |
| 506 | 7.50 | 1.50 | 0.02 | 0.47 |
| 508 | 5.13 | 1.19 | 0.05 | 0.34 |
| 509 | 4.63 | 1.81 | 0.03 | 0.59 |
| 510 | 3.13 | 1.50 | 0.03 | — |
| 544 | 7.50 | 1.00 | 0.03 | 0.47 |
| 544 | 6.75 | 1.50 | 0.09 | 0.47 |
| 546 | 3.25 | 0.94 | 0.04 | 0.47 |
| 546 | 3.25 | 1.19 | 0.05 | 0.23 |
| 613 | 3.25 | 0.31 | 0.05 | 0.47 |
| 613 | 3.25 | 0.56 | 0.05 | 0.47 |
| 614 | 1.00 | 0.38 | 0.02 | 0.13 |
| 615 | 2.94 | 0.75 | 0.05 | 0.23 |
| 616 | 2.94 | 0.56 | 0.05 | — |
| 617 | 6.00 | 0.44 | 0.02 | — |
| 624 | 2.63 | 0.94 | 0.02 | 0.13 |
| 624 | 2.75 | 0.63 | 0.02 | 0.23 |

*Units : $\mu g\ l^{-1}$*

*Recovery and Precision obtained from Eleven Samples (KORDI)*

| Element | Concentration ($\mu g\ l^{-1}$) | | Recovery (%) | RSD (%) |
|---|---|---|---|---|
| | added | found | | |
| Cd | 1.25 | 1.22 ± 0.01 | 98 | 1.0 |
| Co | 2.5 | 2.51 ± 0.06 | 100 | 2.4 |
| Cu | 2.5 | 2.48 ± 0.03 | 99 | 1.3 |
| Ni | 2.5 | 2.50 ± 0.07 | 100 | 2.9 |
| Pb | 2.5 | 2.56 ± 0.05 | 102 | 2.0 |
| Zn | 5.0 | 4.76 ± 0.14 | 96 | 2.9 |

# MARINE RESEARCH INSTITUTE

ADDRESS : Marine Research Institute
Reykjavik
Iceland

ACRONYM : MRI

| Sample No | Hg | Sample No | Hg |
|---|---|---|---|
| 800 513 | 6.0 | 800 581 | 10.8 |
| 800 513 | 7.4 | 800 581 | 4.1 |
| 800 513 | 5.6 | 800 581 | 3.7 |
| 800 513 | 6.9 | 800 581 (fractured) | 10.8 |
| 800 514 | 2.9 | 800 582 | 5.6 |
| 800 514 | 5.8 | 800 583 | 3.1 |
| 800 514 | 3.1 | 800 584 | 4.0 |
| 800 514 | 6.2 | 800 585 | broken |
| 800 516 | 3.1 | 800 623 | 4.1 |

| Sample No | Hg | Sample No | Hg |
|---|---|---|---|
| 800 518 | 2.8 | 800 623 | 3.9 |
| 800 519 | broken | 800 627 | 5.4 |
| 800 520 | 4.1 | 800 627 | 5.1 |
| 800 543 | 3.5 | 800 630 | 5.2 |
| 800 543 | 4.9 | 800 630 | 9.0 |
| 800 543 | 3.0 | 800 632 | 14.7 |
| 800 543 | broken | | |
| 800 545 | 7.4 | | |
| 800 545 | 15.9 | | |
| 800 545 | 6.9 | | |
| 800 545 | broken | | |

*Units : ng $l^{-1}$*

*A single determination of the mercury concentration, without any pretreatment, was performed on each sample. Sample volumes ranged from 265 ml to 425 ml. Estimated precision on each determination ± 0.4 (ng $l^{-1}$).*

| Sample | Cd | Cu | Mn | Fe | Ni | Fb |
|---|---|---|---|---|---|---|
| 800 513 a | 0.030 | 0.062 | 0.008 | 0.278 | 0.261 | 0.095 |
| b | 0.031 | 0.073 | 0.009 | 0.276 | 0.262 | 0.080 |
| c | 0.029 | 0.170 | 0.025 | 0.319 | 0.304 | 0.130 |
| d | 0.022 | 0.098 | 0.006 | 0.314 | 0.265 | 0.084 |
| 514 a | 0.034 | 0.063 | 0.025 | 0.215 | 0.218 | 0.085 |
| b | 0.026 | 0.067 | 0.027 | 0.299 | 0.247 | 0.087 |
| c | 0.030 | 0.085 | 0.017 | 0.410 | 0.245 | 0.158 |
| d | | Chelex resin leaked off the column | | | | |
| 516 | 0.033 | 0.143 | 0.009 | 0.344 | 0.462 | 0.068 |
| 518 | 0.025 | 0.067 | 0.006 | 0.282 | 0.487 | 0.078 |
| 519 | 0.008 | 2.77 | 0.022 | 0.254 | 0.491 | 0.128 |
| 520 | 0.118 | 0.702 | 0.031 | 0.507 | 3.03 | 0.159 |
| 543 | 0.025 | 0.065 | 0.007 | 0.205 | 0.230 | 0.137 |
| b | 0.025 | 0.059 | 0.006 | 0.227 | 0.259 | 0.130 |
| c | 0.010 | 0.062 | 0.005 | 0.254 | 0.229 | 0.127 |
| d | 0.025 | 0.068 | 0.009 | 0.231 | 0.243 | 0.144 |
| 545 a | 0.048 | 0.134 | 0.007 | 0.611 | 0.248 | 0.174 |
| b | 0.047 | 0.156 | 0.015 | 0.636 | 0.224 | 0.123 |
| c | 0.065 | 0.135 | 0.007 | 0.578 | 0.227 | 0.006 |
| d | 0.087 | 0.144 | 0.006 | 0.861 | 0.242 | 0.162 |
| 581 a | 0.026 | 0.064 | 0.009 | 0.206 | 0.243 | 0.156 |
| b | 0.023 | 0.062 | 0.006 | 0.173 | 0.222 | 0.17 |
| d | 0.022 | 0.059 | 0.006 | 0.235 | 0.184 | 0.157 |
| c | 0.023 | 0.065 | 0.009 | 0.216 | 0.234 | 0.233 |
| 582 | 0.036 | 0.293 | 0.007 | 0.349 | 0.386 | 0.110 |
| 583 | 0.035 | 0.268 | 0.007 | 0.265 | 0.515 | 0.137 |
| 584 | 0.034 | 0.097 | 0.012 | 0.317 | 0.533 | 0.019 |
| 585 | 0.034 | 0.093 | 0.008 | 0.250 | 0.479 | 0.072 |
| 800 623 a | 0.011 | 0.176 | 0.024 | 0.787 | 0.128 | 0.254 |
| b | 0.011 | 0.178 | 0.023 | 0.716 | 0.147 | 0.228 |
| 626 a | 0.015 | 0.179 | 0.019 | 0.751 | 0.143 | 0.279 |
| b | 0.013 | 0.192 | 0.020 | 0.762 | 0.150 | 0.217 |
| 627 a | 0.017 | 0.225 | 0.022 | 0.796 | 0.385 | 0.238 |
| b | 0.012 | 0.233 | 0.024 | 0.790 | 0.231 | 0.205 |
| 629 a | 0.016 | 0.171 | 0.026 | 0.727 | 0.120 | 0.246 |
| b | 0.015 | 0.171 | 0.024 | 0.765 | 0.187 | 0.232 |
| 630 a | 0.019 | 0.242 | 0.024 | 0.720 | 0.941 | 0.223 |
| b | 0.019 | 0.241 | 0.028 | 0.739 | 0.965 | 0.239 |
| 631 a | 0.013 | 0.163 | 0.024 | 0.659 | 0.119 | 0.223 |
| b | 0.014 | 0.190 | 0.023 | 0.690 | 0.114 | 0.231 |
| 632 | 0.014 | 0.433 | 0.024 | 0.602 | 0.993 | 0.233 |

*Units : µg $l^{-1}$*

# NETHERLANDS INSTITUTE FOR SEA RESEARCH

ADDRESS : Netherlands Institute for Sea Research
Postbox 59
Den Burg — Texel
The Netherlands

ACRONYM : NISR

| Sample | Element | | | | | |
|---|---|---|---|---|---|---|
| | Cd | | Pb | Cu | Ni | Zn |
| | AAS | ASV | ASV | ASV | AAS | AAS |
| 800525 | 20 | 20 | 0.10 | 0.10 | 0.24 | 1.4 |
| | 15 | 24 | 0.12 | 0.17 | 0.13 | 1.4 |
| | 12 | 20 | 0.15 | 0.14 | 0.13 | 0.6 |
| | 20 | 24 | 0.14 | 0.12 | 0.14 | 1.4 |
| 800526 | 10 | 19 | 0.06 | 0.12 | 0.10 | 1.6 |
| | 18 | 21 | 0.03 | 0.15 | 0.10 | 0.6 |
| | 15 | 21 | 0.03 | 0.16 | 0.12 | 04 |
| | 12 | 20 | 0.03 | 0.21 | 0.10 | 0.4 |
| 800527 | 10 | 22 | 0.10 | 0.08 | 0.10 | 0.4 |
| | 10 | 21 | 0.08 | 0.12 | 0.12 | 0.4 |
| | 30 | 23 | 0.14 | 0.13 | 0.22 | 0.5 |
| | 20 | 23 | 0.10 | 0.15 | 0.12 | 0.6 |
| 800544 | 13 | 19 | 0.04 | 0.18 | 0.12 | 0.4 |
| | 17 | 20 | 0.04 | 0.20 | 0.13 | 0.7 |
| | 22 | 21 | 0.04 | 0.17 | 0.12 | 0.7 |
| | 12 | 21 | 0.04 | 0.10 | 0.12 | 0.4 |
| 800546 | 24 | 19 | 0.05 | 0.30 | 0.12 | 1.4 |
| | * | 19 | 0.08 | 0.23 | * | * |
| | 22 | 21 | 0.08 | 0.35 | 0.10 | 1.4 |
| | 13 | 21 | 0.10 | 0.23 | 0.12 | 1.4 |
| 800596 | 16 | 22 | 0.01 | 0.08 | 0.22 | 1.3 |
| 800597 | 16 | 21 | 0.01 | 0.18 | 0.22 | 1.3 |
| 800598 | 22 | 21 | 0.05 | 0.25 | 0.16 | 1.4 |
| 800600 | 30 | 21 | 0.06 | 0.11 | 0.14 | 2.1 |
| 800601 | 17 | 20 | 0.04 | 0.04 | 0.10 | 1.2 |
| | 35 | 20 | 0.04 | 0.05 | 0.10 | 1.2 |
| | 32 | 19 | 0.04 | 0.05 | 0.10 | 1.2 |
| | 17 | 18 | 0.04 | 0.04 | 0.10 | 1.2 |

*Units : ng $l^{-1}$ for cadmium and $\mu g$ $l^{-1}$ for the other elements.*
** bottle insufficiently filled.*

# NATIONAL INSTITUTE FOR ENVIRONMENTAL STUDIES

ADDRESS : National Institute for Environmental Studies
Yatabe, Ibaraki
Japan

ACRONYM : NIES

**ELEMENT : Cd**

| Sample Code Number | 1 | 2 | 3 | 4 | 5 | 6 | Average | Stnd. Devn. |
|---|---|---|---|---|---|---|---|---|
| 800618,19,20 | 0.028 | 0.032 | 0.036 | spike | spike | spike | 0.032 | ±0.004 |
| 800618 | 0.036 | 0.036 | 0.036 | 0.040 | 0.036 | 0.040 | 0.037 | ±0.002 |
| 800619 | 0.036 | 0.044 | 0.036 | 0.040 | 0.040 | 0.044 | 0.040 | ±0.004 |
| 800620 | 0.036 | 0.024 | 0.036 | 0.028 | 0.024 | 0.036 | 0.031 | ±0.006 |
| 800621 | 0.012 | 0.012 | 0.020 | 0.020 | 0.020 | 0.016 | 0.017 | ±0.004 |
| 800621 | 0.024 | 0.020 | 0.032 | 0.024 | 0.036 | 0.024 | 0.027 | ±0.006 |
| 800555 | 0.028 | 0.028 | 0.028 | spike | spike | spike | 0.028 | ±0.000 |
| 800555 | 0.020 | 0.020 | 0.024 | 0.020 | 0.024 | 0.028 | 0.023 | ±0.003 |
| 800555 | 0.036 | 0.036 | 0.032 | 0.032 | 0.024 | 0.032 | 0.032 | ±0.004 |
| 800557 | 0.032 | 0.032 | 0.024 | 0.028 | 0.040 | 0.040 | 0.033 | ±0.006 |
| 800557 | 0.032 | 0.040 | 0.028 | 0.024 | 0.028 | 0.032 | 0.032 | ±0.003 |
| 800557 | 0.032 | 0.032 | 0.036 | 0.032 | 0.032 | 0.028 | 0.032 | ±0.003 |
| 800535 | 0.036 | 0.040 | 0.036 | 0.048 | 0.044 | 0.044 | 0.041 | ±0.005 |
| 800535 | 0.040 | 0.036 | 0.040 | 0.040 | 0.032 | 0.040 | 0.038 | ±0.003 |
| 800536 | 0.036 | 0.040 | 0.036 | 0.036 | 0.036 | 0.048 | 0.039 | ±0.005 |
| 800536 | 0.032 | 0.032 | 0.032 | 0.040 | 0.040 | 0.036 | 0.035 | ±0.004 |
| 800537 | 0.036 | 0.044 | 0.040 | spike | spike | spike | 0.040 | ±0.004 |
| 800537 | 0.033 | 0.033 | 0.037 | 0.040 | 0.036 | 0.040 | 0.037 | ±0.003 |

*Units : $\mu g\ l^{1}$*

**ELEMENT : Zn**

| Sample Code Number | 1 | 2 | 3 | 4 | 5 | 6 | Average | Stnd. Devn. |
|---|---|---|---|---|---|---|---|---|
| 800618,19,20 | 1.69 | 1.68 | 1.69 | spike | spike | spike | 1.69 | ±0.01 |
| 800618 | 1.65 | 1.67 | 1.65 | 1.63 | 1.64 | 1.65 | 1.65 | ±0.01 |
| 800619 | 1.80 | 1.80 | 1.81 | 1.88 | 1.87 | 1.87 | 1.84 | ±0.04 |
| 800620 | 0.216 | 0.220 | 0.220 | 0.220 | 0.220 | 0.220 | 0.219 | ±0.002 |
| 800621 | 0.064 | 0.056 | 0.064 | 0.060 | 0.060 | 0.068 | 0.062 | ±0.004 |
| 800621 | 0.064 | 0.060 | 0.060 | 0.060 | 0.060 | 0.060 | 0.061 | ±0.002 |
| 800555 | 0.132 | 0.128 | 0.132 | spike | spike | spike | 0.131 | ±0.002 |
| 800555 | 0.112 | 0.116 | 0.112 | 0.112 | 0.112 | 0.112 | 0.113 | ±0.002 |
| 800555 | 0.128 | 0.128 | 0.124 | 0.136 | 0.136 | 0.132 | 0.131 | ±0.005 |
| 800557 | 0.348 | 0.352 | 0.356 | 0.348 | 0.348 | 0.352 | 0.351 | ±0.003 |
| 800557 | 0.244 | 0.240 | 0.240 | 0.232 | 0.240 | 0.232 | 0.238 | ±0.005 |
| 800557 | 0.308 | 0.312 | 0.308 | 0.328 | 0.332 | 0.336 | 0.321 | ±0.013 |
| 800535 | 0.408 | 0.412 | 0.404 | 0.444 | 0.440 | 0.444 | 0.425 | ±0.019 |
| 800535 | 0.204 | 0.208 | 0.204 | 0.208 | 0.208 | 0.208 | 0.207 | ±0.002 |
| 800536 | 0.144 | 0.152 | 0.144 | 0.148 | 0.148 | 0.148 | 0.147 | ±0.003 |
| 800536 | 0.216 | 0.220 | 0.216 | 0.220 | 0.220 | 0.220 | 0.219 | ±0.002 |
| 800537 | 0.244 | 0.236 | 0.244 | spike | spike | spike | 0.241 | ±0.005 |
| 800537 | 0.418 | 0.426 | 0.418 | 0.430 | 0.426 | 0.430 | 0.425 | ±0.005 |

*Units : $\mu g\ l^{1}$*

**ELEMENT : V**

| Sample Code Number | 1 | 2 | 3 | 4 | 5 | 6 | Average | Stnd. Devn. |
|---|---|---|---|---|---|---|---|---|
| 800618,19,20 | 1.36 | 1.37 | 1.36 | spike | spike | spike | 1.36 | ±0.01 |
| 800618 | 1.30 | 1.32 | 1.31 | 1.28 | 1.29 | 1.28 | 1.30 | ±0.02 |
| 800619 | 1.26 | 1.26 | 1.26 | 1.26 | 1.27 | 1.27 | 1.26 | ±0.01 |
| 800620 | 1.29 | 1.29 | 1.29 | 1.24 | 1.25 | 1.24 | 1.27 | ±0.03 |
| 800621 | 1.45 | 1.44 | 1.47 | 1.43 | 1.42 | 1.42 | 1.44 | ±0.02 |
| 800621 | 1.31 | 1.31 | 1.31 | 1.36 | 1.36 | 1.36 | 1.34 | ±0.03 |
| 800555 | 1.45 | 1.46 | 1.44 | spike | spike | spike | 1.45 | ±0.01 |
| 800555 | 1.38 | 1.39 | 1.39 | 1.40 | 1.41 | 1.41 | 1.40 | ±0.01 |
| 800555 | 1.37 | 1.39 | 1.38 | 1.42 | 1.38 | 1.41 | 1.39 | ±0.02 |
| 800557 | 1.41 | 1.42 | 1.42 | 1.40 | 1.40 | 1.38 | 1.41 | ±0.02 |
| 800557 | 1.32 | 1.34 | 1.34 | 1.38 | 1.38 | 1.40 | 1.36 | ±0.03 |
| 800557 | 1.31 | 1.32 | 1.32 | 1.36 | 1.36 | 1.36 | 1.34 | ±0.02 |
| 800535 | 1.23 | 1.25 | 1.24 | 1.22 | 1.23 | 1.23 | 1.23 | ±0.01 |
| 800535 | 1.42 | 1.42 | 1.43 | 1.41 | 1.42 | 1.41 | 1.42 | ±0.01 |
| 800536 | 1.33 | 1.31 | 1.32 | 1.39 | 1.38 | 1.38 | 1.35 | ±0.04 |
| 800536 | 1.43 | 1.45 | 1.43 | 1.42 | 1.44 | 1.44 | 1.44 | ±0.01 |
| 800537 | 1.40 | 1.41 | 1.40 | spike | spike | spike | 1.40 | ±0.01 |
| 800537 | 1.46 | 1.45 | 1.45 | 1.52 | 1.54 | 1.54 | 1.49 | ±0.04 |

*Units : $\mu g\ l^{1}$*

**ELEMENT : Ni**

| Sample Code Number | 1 | 2 | 3 | 4 | 5 | 6 | Average | Stnd. Devn. |
|---|---|---|---|---|---|---|---|---|
| 800618,19,20 | 0.412 | 0.372 | 0.424 | spike | spike | spike | 0.403 | ±0.027 |
| 800618 | 0.256 | 0.288 | 0.288 | 0.276 | 0.316 | 0.296 | 0.287 | ±0.018 |
| 800619 | 0.376 | 0.424 | 0.384 | 0.396 | 0.436 | 0.396 | 0.402 | ±0.023 |
| 800620 | 0.356 | 0.332 | 0.320 | 0.340 | 0.324 | 0.320 | 0.332 | ±0.014 |
| 800621 | 0.164 | 0.172 | 0.204 | 0.204 | 0.152 | 0.192 | 0.181 | ±0.022 |
| 800621 | 0.100 | 0.120 | 0.112 | 0.140 | 0.160 | 0.188 | 0.137 | ±0.033 |
| 800555 | 0.168 | 0.224 | 0.184 | spike | spike | spike | 0.192 | ±0.029 |
| 800555 | 0.228 | 0.244 | 0.200 | 0.272 | 0.240 | 0.212 | 0.233 | ±0.025 |
| 800555 | 0.220 | 0.264 | 0.220 | 0.172 | 0.224 | 0.224 | 0.221 | ±0.029 |
| 800557 | 0.244 | 0.204 | 0.212 | 0.256 | 0.256 | 0.256 | 0.238 | ±0.024 |
| 800557 | 0.256 | 0.268 | 0.240 | 0.256 | 0.244 | 0.232 | 0.249 | ±0.013 |
| 800557 | 0.208 | 0.236 | 0.200 | 0.256 | 0.228 | 0.240 | 0.228 | ±0.021 |
| 800535 | 0.236 | 0.212 | 0.244 | 0.256 | 0.224 | 0.244 | 0.236 | ±0.016 |
| 800535 | 0.212 | 0.236 | 0.284 | 0.260 | 0.232 | 0.200 | 0.237 | ±0.031 |
| 800536 | 0.260 | 0.212 | 0.264 | 0.232 | 0.220 | 0.280 | 0.245 | ±0.027 |
| 800536 | 0.208 | 0.228 | 0.228 | 0.240 | 0.184 | 0.228 | 0.219 | ±0.020 |
| 800537 | 0.264 | 0.228 | 0.204 | spike | spike | spike | 0.232 | ±0.030 |
| 800537 | 0.207 | 0.248 | 0.240 | 0.211 | 0.238 | 0.211 | 0.226 | ±0.018 |

*Units : $\mu g\ l^{1}$*

**ELEMENT : Fe**

| Sample Code Number | 1 | 2 | 3 | 4 | 5 | 6 | Average | Stnd. Devn. |
|---|---|---|---|---|---|---|---|---|
| 800618,19,20 | 0.288 | 0.292 | 0.296 | spike | spike | spike | 0.292 | ±0.004 |
| 800618 | 0.252 | 0.256 | 0.260 | 0.244 | 0.252 | 0.240 | 0.251 | ±0.007 |
| 800619 | 0.336 | 0.352 | 0.348 | 0.424 | 0.440 | 0.420 | 0.387 | ±0.046 |
| 800620 | 0.172 | 0.168 | 0.172 | 0.168 | 0.172 | 0.168 | 0.170 | ±0.002 |
| 800621 | 0.156 | 0.164 | 0.156 | 0.160 | 0.160 | 0.160 | 0.159 | ±0.003 |
| 800621 | 0.116 | 0.112 | 0.108 | 0.260 | 0.260 | 0.248 | 0.184 | ±0.079 |
| 800555 | 0.132 | 0.128 | 0.144 | spike | spike | spike | 0.135 | ±0.008 |
| 800555 | 0.140 | 0.148 | 0.136 | 0.128 | 0.132 | 0.148 | 0.139 | ±0.008 |
| 800555 | 0.164 | 0.164 | 0.164 | 0.212 | 0.208 | 0.216 | 0.188 | ±0.026 |
| 800557 | 0.156 | 1.160 | 0.172 | 0.320 | 0.316 | 0.320 | 0.241 | ±0.086 |
| 800557 | 0.192 | 0.176 | 0.184 | 0.168 | 0.176 | 0.172 | 0.178 | ±0.009 |
| 800557 | 0.180 | 0.180 | 0.184 | 0.300 | 0.312 | 0.308 | 0.244 | ±0.069 |
| 800535 | 0.312 | 0.320 | 0.320 | 0.348 | 0.360 | 0.364 | 0.337 | ±0.023 |
| 800535 | 0.196 | 0.204 | 0.204 | 0.204 | 0.192 | 0.200 | 0.200 | ±0.005 |
| 800536 | 0.196 | 0.212 | 0.212 | 0.312 | 0.304 | 0.304 | 0.257 | ±0.055 |
| 800536 | 0.220 | 0.208 | 0.228 | 0.192 | 0.196 | 0.196 | 0.207 | ±0.015 |
| 800537 | 0.272 | 0.272 | 0.272 | spike | spike | spike | 0.272 | ±0.000 |
| 800537 | 0.381 | 0.406 | 0.389 | 0.408 | 0.408 | 0.390 | 0.397 | ±0.012 |

*Units : μg $l^{1}$*

**ELEMENT : Cu**

| Sample Code Number | 1 | 2 | 3 | 4 | 5 | 6 | Average | Stnd. Devn. |
|---|---|---|---|---|---|---|---|---|
| 800618,19,20 | 0.960 | 0.964 | 0.964 | spike | spike | spike | 0.963 | ±0.002 |
| 800618 | 0.416 | 0.428 | 0.428 | 0.412 | 0.416 | 0.412 | 0.419 | ±0.007 |
| 800619 | 1.33 | 1.33 | 1.33 | 1.38 | 1.38 | 1.39 | 1.36 | ±0.03 |
| 800620 | 0.072 | 0.072 | 0.072 | 0.064 | 0.064 | 0.072 | 0.069 | ±0.004 |
| 800621 | 0.076 | 0.076 | 0.076 | 0.072 | 0.072 | 0.072 | 0.074 | ±0.002 |
| 800621 | 0.064 | 0.064 | 0.064 | 0.072 | 0.064 | 0.064 | 0.065 | ±0.003 |
| 800555 | 0.072 | 0.072 | 0.072 | spike | spike | spike | 0.072 | ±0.000 |
| 800555 | 0.072 | 0.072 | 0.072 | 0.080 | 0.076 | 0.080 | 0.075 | ±0.004 |
| 800555 | 0.072 | 0.072 | 0.072 | 0.072 | 0.072 | 0.072 | 0.073 | ±0.002 |
| 800557 | 0.080 | 0.080 | 0.080 | 0.080 | 0.080 | 0.084 | 0.081 | ±0.002 |
| 800557 | 0.088 | 0.088 | 0.092 | 0.080 | 0.088 | 0.088 | 0.087 | ±0.004 |
| 800557 | 0.080 | 0.080 | 0.080 | 0.084 | 0.080 | 0.084 | 0.081 | ±0.002 |
| 800535 | 0.148 | 0.152 | 0.152 | 0.160 | 0.168 | 0.168 | 0.158 | ±0.009 |
| 800535 | 0.084 | 0.084 | 0.080 | 0.084 | 0.084 | 0.088 | 0.084 | ±0.003 |
| 800536 | 0.096 | 0.100 | 0.100 | 0.096 | 0.100 | 0.104 | 0.099 | ±0.003 |
| 800536 | 0.136 | 0.144 | 0.136 | 0.140 | 0.144 | 0.148 | 0.141 | ±0.005 |
| 800537 | 0.084 | 0.076 | 0.076 | spike | spike | spike | 0.079 | ±0.005 |
| 800537 | 0.136 | 0.136 | 0.136 | 0.148 | 0.148 | 0.143 | 0.141 | ±0.006 |

*Units : μg $l^{1}$*

**ELEMENT : Mo**

| Sample Code Number | 1 | 2 | 3 | 4 | 5 | 6 | Average | Stnd. Devn. |
|---|---|---|---|---|---|---|---|---|
| 800618,19,20 | 7.56 | 7.56 | 7.64 | 7.76 | 7.72 | 7.76 | 7.67 | ±0.09 |
| 800618 | 7.48 | 7.56 | 7.56 | 7.52 | 7.52 | 7.48 | 7.52 | ±0.04 |
| 800619 | 7.28 | 7.32 | 7.28 | 7.76 | 7.84 | 7.76 | 7.54 | ±0.27 |
| 800620 | 8.00 | 7.96 | 8.04 | 7.76 | 7.80 | 7.80 | 7.89 | ±0.12 |
| 800621 | 8.23 | 8.36 | 8.40 | 8.32 | 8.32 | 8.40 | 8.35 | ±0.04 |
| 800621 | 7.88 | 8.00 | 7.92 | 8.40 | 8.48 | 8.44 | 8.19 | ±0.28 |
| 800555 | 7.28 | 7.24 | 7.28 | 7.20 | 7.28 | 7.20 | 7.25 | ±0.04 |
| 800555 | 8.04 | 8.00 | 8.04 | 8.00 | 8.00 | 8.01 | 8.02 | ±0.02 |
| 800555 | 7.88 | 7.96 | 7.96 | 8.16 | 8.12 | 8.08 | 8.03 | ±0.11 |
| 800557 | 7.68 | 7.64 | 7.72 | 7.80 | 7.80 | 7.76 | 7.73 | ±0.07 |
| 800557 | 8.12 | 8.28 | 8.20 | 8.32 | 8.36 | 8.44 | 8.29 | ±0.11 |
| 800557 | 7.88 | 7.92 | 7.92 | 8.24 | 8.24 | 8.28 | 8.08 | ±0.19 |
| 800535 | 7.48 | 7.52 | 7.52 | 8.00 | 8.04 | 8.04 | 7.77 | ±0.29 |
| 800535 | 5.84 | 5.76 | 5.76 | 8.76 | 8.76 | 8.72 | 7.27 | ±1.6 |
| 800536 | 8.08 | 8.04 | 8.08 | 8.44 | 8.40 | 8.48 | 8.25 | ±0.21 |
| 800536 | 8.32 | 8.40 | 8.32 | 8.80 | 8.76 | 8.84 | 8.57 | ±0.25 |
| 800537 | 7.96 | 8.00 | 8.04 | 8.32 | 8.32 | 8.32 | 8.16 | ±0.18 |
| 800537 | 8.82 | 8.82 | 8.69 | 8.97 | 9.01 | 8.97 | 8.88 | ±0.12 |

*Units : µg l⁻¹*

**NOTES**

1. *Sample *800537 was not acidified at Bermuda (see p. 85).*
2. *Two 500 g aliquots were used for analysis*
   *Results — 123 correspond to Ist aliquot*
   *— 456 correspond to 2nd aliquot*
   *Duplicate values were in good agreement except for iron.*
3. *Additional information is given on the following pages.*

## Method Blank

Method blanks were determined by performing total procedure in absence of sea water sample. The blank values were not detected by ICP, except for Fe, and furnace AA was used for quantification in the case of Ni, Cu, Zn and Cd.

*Fe* The high and variable blank values (0.14 ± 0.010 (µg ml$^{-1}$) prevented blank subtraction being meaningful. The reliability of the Fe data is also questionable since agreement for duplicate analyses was not always obtained.

*Ni* The maximum blank by AA accounted for 20 % of the lowest sample concentration (by AA and ICP). As Ni recovery may only be about 90 %, blank concentration was not performed.

*Cu* Blank as determined by AA not significant (< 10 % of lowest sample concentration).

*Zn* Blank as determined by AA not significant (< 10 % of lowest sample concentration).

*Cd* Blank as determined by AA not significant (< 10 % of lowest sample concentration).

## Spike Recovery

500 g aliquots of samples 800618, 19, 20, 800555 and 800537 were spiked with 0.5 ml of multisolution (V 2 µg ml$^{-1}$, Zn 2 µg ml$^{-1}$, Fe 1 µg ml$^{-1}$, Cu 0.5 µg ml$^{-1}$, Ni 0.5 µg ml$^{-1}$, Cd 0.05 µg ml$^{-1}$ in 0.5 % $HNO_3$).

## Recovery Results (%)

| Element | Sample Code Number 0618,19,20 | 0555 | 0537 |
|---|---|---|---|
| V | 105 | 100 | 98 |
| Ni | 93 | 88 | 90 |
| Cd | 88 | 105 | 94 |
| Cu | 101 | 92 | 89 |
| Fe | 98 | 89 | 90 |
| Zn | 96 | 90 | 88 |

*A previous experiment indicated that Mo recovery was about 80 %.*

## Method Precision

8 aliquots (500 ml) of Japan Sea sample were analysed in an identical manner as the Bermuda samples.

| Element | Concentration (µg l$^{-1}$) | RSD |
|---|---|---|
| Mo | 7.76 | 5 |
| V | 1.46 | 1 |
| Ni | 0.236 | 10 |
| Cd | 0.016 | 31 |
| Cu | 0.308 | 6 |
| Fe | 1.26 | 4 |
| Zn | 1.42 | 2 |

## Instrument Measuring Precision and Detection Limits

| Element | RSD[1] | Concentration $\mu g\ ml^{-1}$ | Detection Limit[2] $\mu g\ ml^{-1}$ |
|---|---|---|---|
| Mo | 1.2 | 1.0 | 0.004 |
| V | 1.3 | 0.5 | 0.003 |
| Ni | 5.8 | 0.1 | 0.008 |
| Cd | 12 | 0.01 | 0.001 |
| Cu | 2.0 | 0.1 | 0.001 |
| Fe | 1.5 | 0.2 | 0.003 |
| Zn | 0.9 | 0.2 | 0.001 |

*1. calculated from data for check standard.*

*2. calculated (as 2 $\sigma$) from data for distilled water (zero concentration).*

## Instrument Operation

Standardisation of the instrument was repeated for each pair of sea-water extracts to minimize effects of instrumental drift.

The sequence outlined below was repeated 18 times (18 samples) to generate the data base ;

(1) Instrument standardized with 1 ppm multisolution (Mo, V, Ni, Cd, Cu, Fe, Zn, Na in 0.5 % $HNO_3$ — high standard) and distilled water (low standard).

(2) Distilled water analysed — 3 measurements : results enabled detection limits to be calculated.

(3) Sea Water extract (from first 500 g aliquot ; e.g., 0618, 19, 20 — A) analysed — 3 measurements.

(4) Sea water extract (from second 500 g aliquot ; e.g., 0618, 29, 20 — B) analysed — 3 measurements.

(5) Check standard (Mo 1 $\mu g\ ml^{-1}$, V 0.5 $\mu g\ ml^{-1}$, Ni 0.1 $\mu g\ ml^{-1}$, Cd 0.01 $\mu g\ ml^{-1}$, Cu 0.1 $\mu g\ ml^{-1}$, Fe 0.2 $\mu g\ ml^{-1}$, Zn 0.2 $\mu g\ ml^{-1}$ in 0.5 % $HNO_3$) analysed : results enabled instrument measuring precision to be calculated.

One day before preconcentration the pH's of the samples were measured to check that acidification on the ship had been performed. For this, about 5 ml of sea water were transferred to a small beaker for measurement. It was found that one sample — *800537 — had not been acidified (pH 7.46, see table). This sample was subsequently acidified to pH 1.83 with 1 ml $HNO_3$ (subboiling distilled).

| Sample Code No. | pH |
|---|---|
| 00618,19,20 | 1.73 |
| 00618 | 1.74 |
| 00619 | 1.74 |
| 00620 | 1.74 |
| 00621 | 1.75 |
| 00621 | 1.77 |
| 00555 | 2.02 |
| 00555 | 1.88 |
| 00555 | 1.93 |
| 00557 | 1.94 |
| 00557 | 1.85 |
| 00557 | 1.84 |
| 00535 | 1.79 |
| 00535 | 1.80 |
| 00536 | 1.79 |
| 00536 | 1.85 |
| *800537 | 7.46 (1.83) |
| 00537 | 1.83 |

# NAVAL RESEARCH LABORATORY

ADDRESS : Naval Research Laboratory
Code 4330
Washington, D.C. 20375
U.S.A.

ACRONYM : NRL

| Yellow Tag | tef | Cd | Fe | Zn | Cu | Ni |
|---|---|---|---|---|---|---|
| St. Georges West Surface Water | | | | | | |
| 800700 | 146 | 2.0 | 5.4 | 0.6 | 0.2 | 1.7 |
| 800700 | 143 | 3.1 | 7.7 | 0.5 | 0.6 | 2.3 |
| 800700 | 132 | 1.8 | 5.4 | 0.6 | 0.2 | 1.7 |
| 800700 | 140 | 3.2 | 4.8 | 1.0 | 0.3 | 2.1 |
| below stored 3 months, acidified, prior to analysis | | | | | | |
| 800700 | 301 | 1.9 | 3.4 | 0.6 | 0.3 | 3.3 |
| 800700 | 303 | 2.8 | 3.4 | 0.6 | 0.3 | 3.5 |
| 800700 | 303 | 0.7 | 5.5 | bdl | 0.8 | 3.1 |
| Panulirus Station Water | | | | | | |
| 800504 | 147 | bdl | 6.1 | bdl | 0.7 | 1.7 |
| 800504 | 159 | 2.0 | 2.4 | 0.7 | 0.07 | 1.3 |
| 800504 | 149 | 2.0 | 1.4 | 0.4 | 0.1 | 1.4 |
| 800504 | 110 | 1.7 | 2.9 | 0.2 | 0.4 | 1.9 |
| 800505 | 138 | 1.7 | 2.8 | 0.05 | 0.05 | 1.6 |
| 800505 | 135 | 0.3 | 4.2 | bdl | 0.8 | 2.1 |
| 800505 | 133 | 0.8 | 5.9 | bdl | 0.4 | 1.6 |
| 800505 | 139 | 1.8 | 3.5 | 0.3 | 0.02 | 1.6 |
| 800506 | 136 | 0.8 | 5.6 | 0.07 | 0.05 | 1.6 |
| 800508 | 137 | 1.1 | 3.5 | bdl | 0.6 | 1.9 |

*Time series for surface water stored in sampling containers. These samples stored acidified for 3 months prior to analysis.*

| | Yellow Tag | tef | Cd | Fe | Zn | Cu | Ni |
|---|---|---|---|---|---|---|---|
| +2 hr | 800629 | 307 | 2.5 | 2.3 | 0.2 | 0.09 | 3.1 |
| | 800629 | 308 | 3.1 | 5.1 | 0.5 | 0.05 | 3.8 |
| +4.5 hr | 800626 | 304 | 0.8 | 0.7 | bdl | 1.2 | 2.8 |
| +9 hr | 800631 | 305 | 1.7 | 5.7 | 0.2 | 0.2 | 3.5 |
| | 800631 | 306 | 3.3 | 2.8 | 0.5 | 0.3 | 3.3 |

| Yellow Tag | tef | Cd | Fe | Zn | Cu | Ni |
|---|---|---|---|---|---|---|
| 800509 | 131 | 3.1 | 3.3 | 7.9 | 0.7 | 2.6 |
| 800544 | 148 | 0.6 | 7.8 | bdl | 1.1 | 2.0 |
| 800544 | 151 | 1.3 | 4.3 | 0.4 | 0.03 | 1.7 |
| 800544 | 150 | 1.4 | 4.2 | 0.2 | bdl | 1.6 |
| 800544 | 142 | bdl | 6.1 | bdl | 1.0 | 1.1 |
| 800546 | 144 | 1.5 | 5.0 | 0.2 | 0.5 | 1.6 |
| 800546 | 141 | 1.3 | 5.8 | bdl | 0.04 | 1.7 |
| 800546 | 134 | 1.2 | 4.9 | 0.2 | 0.3 | 1.9 |
| 800546 | 145 | 1.4 | 7.4 | bdl | bdl | 1.8 |
| 800592 | 128 | 2.3 | 1.1 | 0.8 | 0.2 | 1.5 |
| 800592 | 121 | 2.5 | 1.4 | 0.9 | 0.2 | 1.7 |
| 800592 | 127 | 2.1 | 3.3 | 0.7 | bdl | 1.7 |
| 800592 | 115 | 1.3 | 4.1 | 0.4 | 0.4 | 1.6 |
| 800593 | 129 | 1.8 | 1.8 | bdl | 0.1 | 1.6 |
| 800594 | 106 | 1.4 | 3.5 | 0.2 | 0.1 | 1.9 |
| 800595 | 105 | 2.8 | 1.6 | 1.0 | 0.6 | 2.1 |
| 800596 | 111 | 1.8 | 4.4 | 0.2 | 0.6 | 2.1 |

** All metal values are reported in $\mu g\ l^{1}$ (ppb) on a weight/weight basis. All samples are unfiltered. bdl = below detection limit.*

*The following data are offered for interest only, because the many variables associated with the intercalibration exercise test partially invalidate the artificial grouping.*

Averages for

| | | | | | |
|---|---|---|---|---|---|
| All surface waters | 2.24 | 4.88 | 0.53 | 0.39 | 2.85 |
| All deep waters | 1.60 | 4.01 | 0.41 | 0.39 | 1.74 |

** All metal values are reported in $\mu g\ l^{1}$ (ppb) on a weight/weight basis. All samples are unfiltered. bdl = below detection limit.*

# SAGAMI CHEMICAL RESEARCH CENTER


ADDRESS : Sagami Chemical Research Center
Nishi-Ohnuma 4-4-1
Sagamihara, Kanawaga 229
Japan

ACRONYM : SAGAMI


### Element determined : Hg

| Sample | Bottle | Volume taken | Concentration (ng $l^{-1}$) |
|---|---|---|---|
| 800540 | 500 ml | 300 ml | 0.9 |
| | | 300 | 1.1 |
| 800541 | 500 ml | 300 | 1.0 |
| | | 300 | 1.2 |
| 800542 | 500 ml | 300 | 1.0 |
| 800586 | 500 ml | 300 | 1.0 |
| | | 300 | 1.1 |

*Preconcentration : Au- amalgamation*
*Instrumentation : Cold-vapour atomic absorption spectrometry.*

### Element determined : Cd

| Sample | Bottle | Volume taken | n | Concentration (μg $l^{-1}$) |
|---|---|---|---|---|
| 800540 | 5 1 | 1 1 | 1 | 0.049 |
| 0541 | " | " | 1 | 0.031 |
| 0542 | " | " | 1 | 0.052 |
| 0540 | 1 1 | 500ml | 1 | 1.6 |
| 0541 | " | " | 1 | 0.32 |
| 0542 | " | " | 1 | 0.030 |
| 0587 | " | " | 1 | 0.037 |
| 0589 | " | " | 1 | 0.033 |
| 0590 | " | " | 1 | 0.033 |
| 0590 | " | " | 1 | 0.039 |
| 0590 | " | " | 1 | 0.035 |
| Surface | 2 1 | " | 1 | 0.017 |

*Estimated error : ± 6 %*

### Element determined : Cu

| Sample | Bottle | Volume taken | n | Concentration (μg $l^{-1}$) |
|---|---|---|---|---|
| 800540 | 5 1 | 1 1 | 1 | 0.36 |
| 0541 | " | " | 1 | 0.40 |
| 0542 | " | " | 1 | 0.34 |
| 0540 | 1 1 | 500ml | 1 | 0.33 |
| 0541 | " | " | 1 | 0.69 |
| 0542 | " | " | 1 | 0.33 |
| 0587 | " | " | 1 | 1.40 |
| 0589 | " | " | 1 | 0.28 |
| 0590 | " | " | 1 | 0.35 |
| 0590 | " | " | 1 | 0.30 |
| 0590 | " | " | 1 | 0.31 |
| Surface | 2 1 | " | 1 | 0.58 |

*Estimated error : ± 6 %*

### Element determined : Zn and Pb

| Sample | Bottle | Volume taken | n | Concentration (μg $l^{-1}$) Zn | Pb |
|---|---|---|---|---|---|
| 800540 | 5 1 | 1 1 | 1 | 0.98 | 0.28 |
| 0541 | " | " | 1 | 1.20 | 0.23 |
| 0542 | " | " | 1 | 1.18 | 0.20 |
| 0540 | 1 1 | 500ml | 1 | 1.41 | 0.34 |
| 0541 | " | " | 1 | 1.13 | 0.30 |
| 0542 | " | " | 1 | 1.05 | 0.23 |
| 0587 | " | " | 1 | 2.47 | 0.30 |
| 0589 | " | " | 1 | 0.80 | 0.23 |
| 0590 | " | " | 1 | 0.75 | 0.23 |
| 0590 | " | " | 1 | 0.85 | 0.17 |
| 0590 | " | " | 1 | 0.70 | 0.18 |
| Surface | 2 1 | " | 1 | 1.78 | 0.27 |

*Estimated error : ± 6 %*

### Determination procedure

Element analysed : Cd, Cu, Zn and Pb

### Instruments used :

Cu and Zn : Atomic absorption Spectrometer, Nippon Jarrell-Ash Co., AA-I, Mark II.
Cd and Pb : Carbon rod flameless (HU-10 Flameless Atomizor, Nippon Jarrel-Ash co.) and a.a.s.

### Pretreatment of sample

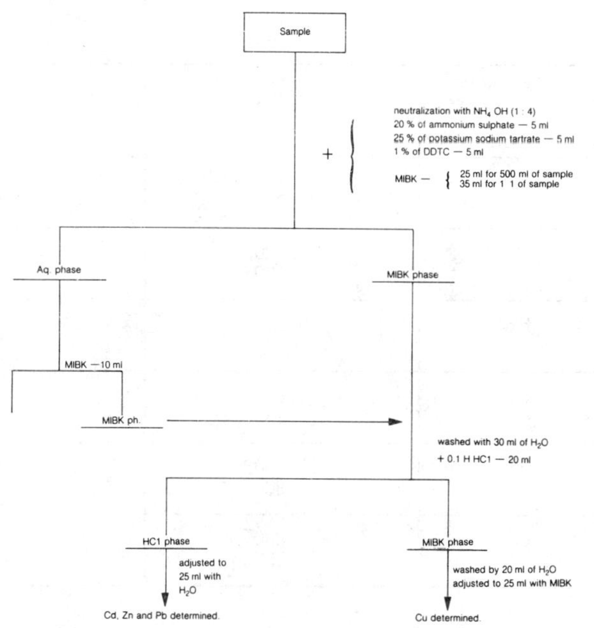


Determination : Cu : 324.7 nm
Zn : 213.9 nm
Flame a.a.s.

Cd : 223.8 nm
Pb : 283.3 nm
Flameless a.a.s.

# SKIDAWAY INSTITUTE OF OCEANOGRAPHY

ADDRESS : Skidaway Institut of Oceanography
Box 13687
Savannah, Georgia 31406
U.S.A.

ACRONYM : SKIDAWAY

| SAMPLE # | COPPER | CADMIUM | NICKEL | ZINC |
|---|---|---|---|---|
| 530 | 88<br>79<br>108<br>108 | 42<br>34<br>41<br>a | ≫1000*<br>340<br>430<br>a | 714<br>571<br>571<br>a |
| 531 | 108<br>122<br>127<br>98 | 19+<br>40<br>31<br>28 | ≫1000*<br>378<br>261<br>365 | 625<br>553<br>428<br>473 |
| 532 | 74<br>144*<br>88<br>88 | 55<br>50<br>43<br>41 | 646<br>664<br>455<br>369+ | 545<br>482<br>446<br>509 |
| 549 | 171<br>117<br>147<br>195 | 66<br>67<br>29<br>20 | ≫1000*<br>336<br>614<br>952 | 419*<br>295<br>280<br>270 |
| 551 | 117<br>88<br>108<br>163* | 20+<br>49<br>47<br>59 | 698*<br>262<br>503<br>293 | 263+<br>670<br>660<br>714 |
| 570 | 135<br>132<br>137 | 51<br>40<br>47 | 518<br>344<br>193 | 638<br>504<br>607 |
| 577* | ≫1000<br>≫1000<br>≫1000 | 24<br>≫100<br>≫100 | 540<br>544<br>790 | 816<br>171<br>≫1000 |
| 578 | 110<br>120<br>140 | 4+<br>26<br>20 | 670<br>718<br>822 | 250<br>230<br>300 |
| 579 | 120<br>89<br>130 | 31<br>41<br>44 | ≫1000*<br>302<br>411 | 180<br>120<br>110 |
| 613 | 170<br>370* | 39<br>35 | 430<br>690 | 54<br>80 |
| 621 | 120<br>100 | 5.4<br>3.1 | 260<br>460 | 72<br>64 |
| 623 | 220<br>260 | 6.2<br>3.9 | 280<br>280 | 300<br>150 |
| 625 | 280<br>240 | 3.9<br>7.7 | 260<br>260 | 200<br>240 |
| 627 | 290<br>300 | a<br>11 | a<br>260 | a<br>870 |
| 628 | 280<br>290 | 11<br>7.7 | 520<br>690 | ≫1000*<br>200 |
| 630 | 340<br>310 | 15<br>7.7 | 1000<br>810 | ≫1000*<br>660 |
| 632 | 400 | 5.4 | 690 | 610 |
| 633 | 320<br>330 | 11<br>6.2 | 340<br>340 | 420<br>320 |

**NOTES :** ** — Sample contamination suspected.*
*a — Sample extract pilled*
*+ — Poor extraction effeciency suspected.*
*Units : ng $l^{-1}$*

# TEXAS A & M UNIVERSITY DATA

ADDRESS : Texas A & M University
College Station
Texas 77843
U.S.A.

ACRONYM : TAMU

| Sample | | Cu | x ± S | RSD $\frac{s}{x}$ (CV) |
|---|---|---|---|---|
| | | | | % |
| 535 | A | 93.8 | 108.8 ± 13.4 | 12.3 |
| | B | 101.3 | | |
| | C | 119.3 | | |
| | D | 120.8 | | |
| 536 | A | 123.0 | 131.6 ± 19.4 | 14.7 |
| | B | 108.3 | | |
| | C | 147.0 | | |
| | D | 148.2 | | |
| 537 | A | 323.7 | 165.5 ± 105.5 | 63.7 |
| | B | 113.3 | | |
| | C | 117.8 | | |
| | D | 107.3 | | |
| 555 | A | 100.6 | 101.7 ± 4.8 | 4.7 |
| | B | 105.8 | | |
| | C | 95.3 | | |
| | D | 105.1 | | |
| 557 | A | 107.3 | 106.5 ± 1.3 | 1.2 |
| | B | 105.1 | | |
| | C | 107.8 | | |
| | D | 105.8 | | |
| 603 | A | 130.5 | 137.3 ± 27.4 | 20.0 |
| | B | 111.8 | | |
| | C | 130.5 | | |
| | D | 176.2 | | |
| 604 | | 319.0 | | |
| 605 | | 200.9 | | |
| 606 | | 129.0 | | |
| 607 | | 131.8 | | |

*Units : ng kg^1*

| Sample | | Cd | x | S | RSD |
|---|---|---|---|---|---|
| | | | | | % |
| 535 | A | 39 | 36.5 | 3.5119 | 9.62 |
| | B | 33 | | | |
| | C | 34 | | | |
| | D | 40 | | | |
| 536 | A | 37 | 46.8 | 9.3229 | 19.94 |
| | B | 43 | | | |
| | C | 48 | | | |
| | D | 59 | | | |
| 537 | A | 62 | 40.3 | 14.5688 | 36.20 |
| | B | 32 | | | |
| | C | 35 | | | |
| | D | 32 | | | |
| 555 | A | 32 | 29.0 | 3.4641 | 11.95 |
| | B | 30 | | | |
| | C | 24 | | | |
| | D | 30 | | | |

| Sample | | Cd | x | S | RSD |
|---|---|---|---|---|---|
| 557 | A | 38 | 38.0 | 0 | 0 |
| | B | 38 | | | |
| | C | 38 | | | |
| | D | 38 | | | |
| 603 | A | 36 | 37.0 | 1.1547 | 3.12 |
| | B | 38 | | | |
| | C | 38 | | | |
| | D | 36 | | | |
| 604 | | 40 | | | |
| 605 | | 35 | | | |
| 606 | | 35 | | | |
| 607 | | 35 | | | |

*Units : ng kg^1*

| Sample | | Ni | x | S | RSD |
|---|---|---|---|---|---|
| | | | | | % |
| 535 | A | 419 | 421 | 9.07 | 2.16 |
| | B | 422 | | | |
| | C | 410 | | | |
| | D | 432 | | | |
| 536 | A | 419 | 393 | 27.21 | 6.93 |
| | B | 365 | | | |
| | C | 413 | | | |
| | D | 374 | | | |
| 537 | A | 480 | 416 | 45.36 | 10.92 |
| | B | 378 | | | |
| | C | 413 | | | |
| | D | 391 | | | |
| 555 | A | 369 | 350 | 34.69 | 9.93 |
| | B | 375 | | | |
| | C | 299 | | | |
| | D | 355 | | | |
| 557 | A | 377 | 365 | 9.03 | 2.47 |
| | B | 367 | | | |
| | C | 361 | | | |
| | D | 356 | | | |
| 603 | A | 368 | 367 | 7.85 | 2.15 |
| | B | 371 | | | |
| | C | 374 | | | |
| | D | 356 | | | |
| 604 | | 554 | | | |
| 605 | | 503 | | | |
| 606 | | 515 | | | |
| 607 | | 470 | | | |

*Units : ng kg^1*

# UNIVERSITY OF CONNECTICUT

ADDRESS : University of Connecticut
Avery Point, Groton
Connecticut 06340
U.S.A.

ACRONYM : UCON

**Concentrations in replicate determinations**

| SAMPLE BOTTLE NO. | | Cd | Cu | Ni | Zn |
|---|---|---|---|---|---|
| 800611 (3 bottles) | 1<br>2<br>3 | .025, .019,<br>.018, .018,<br>.026, .025 | .131, .136,<br>.145, .189<br>.226, .203 | .131, .136,<br>.160, .146,<br>.215, .203 | .400,<br>.245, .313,<br>.335, .350 |
| 800608/609/610 | | .046 .048 | .962, .979 | .305, .300 | .305, .648 |
| 800610 | | .037, .039 | .160, .117 | .209, .218 | .455 |
| 800609 | | .025, .035 | .841, .945 | .205, .215 | .385 |
| 800608 | | 0.34, .068<br>.083 | .620, .713,<br>.733 | .262, .258,<br>.245 | .400, .650,<br>.780 |
| 800623 (algae in bottles) | 1<br>2 | .021, .023<br>.048, .040 | .320, .342<br>.301, .345 | .123, .129<br>.115, .135 | .390, .375<br>.580, .680 |
| 800625 (algae) | | .050 | .258, .455 | .123, .205 | .605 |
| 800628 (algae, sample received incorrectly acidified, pH 6) | | .018, .024 | .174, .133 | .093, .096 | .353, .425 |
| 800633 (algae, pH 5) | | .017, .045 | .236, .214 | .79, .098 | .470, .545 |

*Units : µg kg⁻¹*

## UNIVERSITY OF DELAWARE


ADDRESS : University of Delaware
Newak
Delaware 19711
U.S.A.

ACRONYM : UDEL


| Sample Number | Trace Metal Concentration Ni | Cu | Cd |
|---|---|---|---|
| 800700 bach surface | 182 ± 25) (203 ± 18) | lost (166 ± 33) | 2.4 ± 0.2 (3.2 ± 0.8) |
| 800552 | (265 ± 7 (291 ± 18) | 292 ± 8 (210 ± 15) | 36 ± 3 (13 ± 3) |
| 800526 | 234 ± 11 (265 ± 27) | 134 ± 4 (122 ± 5) | 32 ± 1 (21 ± 8) |
| 800550 | 238 ± 40 (262 ± 20) | 106 ± 13 (103 ± 10) | 23 ± 2 (24 ± 3) |
| 800527 | 238 ± 6 (230 ± 6) | 116 ± 34 (113 ± 10) | 24 ± 6 (25 ± 3) |
| 800529 (592 ?) | 200 ± 20 | 102 ± 28 | 24 ± 5 |
| 800525 | 199 ± 73 | 133 ± 9 | 23 ± 5 |
| 800625 | 41 ± 4 | 178 ± 21 | 4.4 ± 0.6 |
| 800624 | 50 ± 7 | 179 ± 46 | 5.3 ± 2.1 |
| 800628 | 70 ± 10 | 188 ± 30 | 6.2 ± 2.3 |
| 800593 | 200 ± 12 | 240 ± 11 | 15.5 ± 0.4 |
| 800594 | 214 ± 5 | 166 ± 6 | 25.2 ± 0.5 |
| 800305 | 274 ± 16 | 99 ± 12 | 20 ± 2 |
| 800700 batch surface | 182 ± 21 (203 ± 18) | lost (166 ± 33) | 2.4 ± 0.4 (3.2 ± 0. 0.8) |
| 800552 | 265 ± 7 (291 ± 18) | 292 ± 6 (210 ± 15) | 36 ± 3 (13 ± 3) |
| 800526 | 234 ± 11 (265 ± 27) | 134 ± 4 (122 ± 5) | 32 ± 1 (21 ± 8) |
| 800550 | 238 ± 45 (262 ± 20) | 106 ± 13 (103 ± 10) | 23 ± 2 (24 ± 3) |
| 800527 | 238 ± 6 (230 ± 6) | 115 ± 22 (113 ± 10) | 24 ± 6 (25 ± 3) |
| 800592 | 200 ± 20 | 101 ± 28 | 23 ± 5 |
| 800525 | 204 ± 56 | 133 ± 9 | 23 ± 1 |
| 800625 | 41 ± 7 | 178 ± 21 | 4.4 ± 2.4 |
| 800624 | 50 ± 7 | 179 ± 46 | 5.3 ± 2.1 |
| 800628 | 70 ± 14 | 188 ± 30 | 6.2 ± 2.3 |
| 800593 | 200 ± 12 | 240 ± 11 | 15.5 ± 0.4 |
| 800594 | 214 ± 5 | 166 ± 6 | 25.2 ± 0.5 |
| 800595 | 274 ± 16 | 99 ± 12 | 20 ± 2 |

*Units ng $l^{-1}$*
*Values reported are for sea water aliquots returned, processed and analysed at Delaware. Values in brackets were generated at BBS on Bermuda.*

| | BOTTLE | Ni | SD | Cu | SD | Cd | SD |
|---|---|---|---|---|---|---|---|
| 800700 | 1 | 167 | | | | 2.0 | |
| | | 160 | | | | 2.7 | |
| | 2 | 203 | | | | 2.2 | |
| | | 196 | | | | 2.8 | |
| | | 182 | 21 | | | 2.4 | 0.4 |
| 800552 | 1 | 260 | | 291 | | 33.8 | |
| | 2 | 263 | | 298 | | 38.0 | |
| | 3 | 273 | | 287 | | 37.5 | |
| | | 265 | 7 | 292 | 6 | 36 | 3 |
| 800526 | 1 | 227 | | 131 | | 33.3 | |
| | | 242 | | 136 | | 31.5 | |
| | | 234 | 11 | 134 | 4 | 32 | 1 |
| 800550 | 1 | 202 | | 116 | | 24.1 | |
| | | 207 | | 116 | | 24.0 | |
| | 2 | 299 | | 090 | | 22.0 | |
| | | 244 | | 101 | | 21.8 | |
| | | 238 | 45 | 106 | 13 | 23 | 1 |
| 800527 | 1 | 231 | | 101 | | 30.5 | |
| | | 239 | | 092 | | 26.1 | |
| | 2 | 243 | | 142 | | 18.7 | |
| | | | | 123 | | 19.7 | |
| | | 238 | 6 | 115 | 22 | 24 | 6 |
| 800592 | 1 | 196 | | 081 | | 17.6 | |
| | 2 | 182 | | 088 | | 25.2 | |
| | 3 | 222 | | 133 | | 27.6 | |
| | | 200 | 20 | 101 | 28 | 23 | 5 |
| 800525 | 1 | 246 | | 145 | | 22.1 | |
| | | 257 | | 130 | | 24.0 | |
| | 2 | 167 | | 123 | | 23.0 | |
| | | 146 | | 134 | | | |
| | | 204 | 56 | 133 | 9 | 23 | 1 |
| 800625 | 1 | 45 | | 163 | | 2.0 | |
| | | 31 | | 164 | | 7.7 | |
| | 2 | 46 | | 179 | | 3.4 | |
| | | 43 | | 207 | | 4.5 | |
| | | 41 | 7 | 178 | 21 | 4.4 | 2.4 |
| 800624 | 1 | 51 | | 212 | | 3.7 | |
| | | 57 | | 147 | | 4.4 | |
| | 2 | 43 | | (435) | | 7.8 | |
| | | 50 | 7 | 179 | 46 | 5.3 | 2.2 |
| 800628 | 1 | 80 | | 209 | | 7.9 | |
| | | 60 | | 166 | | 4.6 | |
| | | 70 | 14 | 188 | 30 | 6.2 | 2.3 |
| 800593 | 1 | 209 | | 232 | | 15.3 | |
| | | 192 | | 248 | | 15.8 | |
| | | 200 | 12 | 240 | 11 | 15.6 | 0.4 |
| 800594 | 1 | 218 | | 170 | | 24.9 | |
| | | 211 | | 162 | | 25.6 | |
| | | 214 | 5 | 166 | 6 | 25.2 | 0.5 |
| 800595 | 1 | 286 | | 90 | | 18.6 | |
| | | 262 | | 108 | | 21.9 | |
| | | 274 | 17 | 99 | 12 | 20 | 2.3 |

*Units : ng $l^{-1}$*

# UNIVERSITY OF SAINS MALAYSIA DATA

ADDRESS : Universiti Sains Malaysia
Penang
Malaysia

ACRONYM : USM

Trace Metals Content

| Code No. of Water Samples | Cd | Co | Cr | Cu | Fe | Mo | Ni | Pb | Zn |
|---|---|---|---|---|---|---|---|---|---|
| 000607 | 0.0091 | 0.127 | BDL* | 0.0052 | 0.0499 | BDL | 0.1273 | 0.0462 | 0.0043 |
| 000600 | 0.0061 | 0.182 | BDL | 0.0103 | 0.0588 | 0.0003 | 0.1454 | 0.0616 | 0.0043 |
| 000609 | 0.0001 | 0.127 | BDL | 0.0052 | 0.0499 | BDL | 0.1454 | 0.0616 | 0.0043 |
| 000610 | 0.0061 | 0.145 | BDL | 0.0052 | 0.066 | BDL | 0.1273 | 0.0462 | 0.0054 |
| 000611 | 0.0091 | 0.109 | BDL | 0.0052 | 0.00 | BDL | 0.1818 | 0.0693 | 0.0043 |
| 000611 | 0.0001 | 0.164 | BDL | 0.0052 | 0.033 | BDL | 0.1818 | 0.0539 | 0.0043 |
| 000611 | 0.0091 | 0.164 | BDL | 0.0052 | 0.033 | BDL | 0.1818 | 0.0462 | 0.0043 |
| 000611 | 0.0091 | 0.182 | BDL | 0.0062 | 0.0499 | BDL | 0.1454 | 0.0539 | 0.0043 |
| 000525 | 0.0091 | 0.145 | BDL | 0.0103 | 0.0499 | BDL | 0.1454 | 0.0616 | 0.0110 |
| 000525 | 0.0091 | 0.145 | BDL | 0.0103 | 0.0333 | BDL | 0.1454 | 0.0616 | 0.0086 |
| 000525 | 0.0091 | 0.145 | BDL | 0.0103 | 0.0499 | BDL | 0.1454 | 0.0616 | 0.0107 |
| 000525 | 0.0091 | 0.145 | BDL | 0.0103 | 0.0417 | BDL | 0.1454 | 0.0616 | 0.0107 |
| 000525 | 0.0091 | 0.145 | BDL | 0.0103 | 0.0333 | BDL | 0.1454 | 0.0616 | 0.0107 |
| 000525 | 0.0091 | 0.145 | BDL | 0.0103 | 0.0333 | BDL | 0.1454 | 0.0539 | 0.0107 |
| 000526 | 0.0061 | 0.102 | BDL | 0.0062 | 0.0499 | BDL | 0.0909 | 0.0616 | 0.0043 |
| 000526 | 0.0061 | 0.145 | BDL | 0.0062 | 0.0499 | BDL | 0.1454 | 0.0616 | 0.0064 |
| 000526 | 0.0091 | 0.145 | BDL | 0.0103 | 0.0588 | BDL | 0.1454 | 0.0616 | 0.0064 |
| 000527 | 0.0061 | 0.145 | BDL | 0.0052 | 0.0499 | BDL | 0.1091 | 0.0539 | 0.0064 |
| 000527 | 0.0091 | 0.145 | BDL | 0.0105 | 0.0167 | BDL | 0.1273 | 0.0462 | 0.0043 |
| 000527 | 0.0091 | 0.145 | BDL | 0.0103 | 0.0333 | BDL | 0.1273 | 0.0509 | 0.0064 |
| 000543 | 0.0061 | 0.164 | BDL | 0.0103 | 0.0499 | BDL | 0.1818 | 0.0462 | 0.0064 |
| 000543 | 0.0091 | 0.182 | BDL | 0.0515 | 0.0333 | BDL | 0.1454 | 0.0616 | 0.0043 |
| 000543 | 0.0061 | 0.145 | BDL | 0.0103 | 0.0417 | BDL | 0.1091 | 0.0462 | 0.0064 |
| 000543 | 0.0001 | 0.127 | BDL | 0.0103 | 0.0333 | BDL | 0.1454 | 0.0693 | 0.0064 |
| 000545 | 0.0061 | 0.109 | BDL | 0.0103 | 0.0499 | BDL | 0.1273 | 0.0693 | 0.0064 |
| 000545 | 0.0061 | 0.109 | BDL | 0.0103 | 0.0499 | BDL | 0.1454 | 0.0616 | 0.0006 |
| 000545 | 0.0061 | 0.127 | BDL | 0.0103 | 0.0249 | BDL | 0.1273 | 0.0616 | 0.0006 |

** BDL Below detectable level*

*Units : $\mu g\ ml^{-1}$*

# Chapter II
# Organochlorines

K.H. Palmork, Institute of Marine Research, Bergen, Norway
J.C. Duinker, Netherlands Institute for Sea Research, Den Burg, Texel
and
A.H. Knap, Bermuda Biological Station for Research, Inc., Bermuda
J.P. Villeneuve, IAEA International Laboratory of Marine Radioactivity, Monaco

## Abstract

This report presents the results of the sampling of sea water and the analyses of organochlorines in sea water which took place at the Bermuda Biological Station for Research, Inc., 11-26 January 1980. The exercise was designed to evaluate different ways of sampling and intercalibrate the analyses of sea-water samples collected in the same homogeneous water mass off Bermuda. To our knowledge, this exercise represented the first time that temperature-programmed glass-capillary gas chromatography was used in an international intercalibration effort for the analyses at the test site (i.e., Bermuda) and in the home laboratories, and resulted in the resolution of individual PCB components.

The apparatus used, laboratory conditions, analytical instruments and chemicals used are described. The shipboard activities are also presented. The results obtained at Bermuda and in the home laboratories are discussed.

## Introduction

One of the aims of the IOC Global Investigation of Pollution in the Marine Environment (GIPME) and of the UNEP Global Governmental Monitoring System is to delineate temporal trends in the levels of certain organic and inorganic chemicals in the open ocean. To be able to fullfill this proposed international objective a Pilot Project was designed, as described in the section on BACKGROUND.

At the Second Session of the IOC, GIPME Group of Experts on Methods, Standards and Intercalibration (GEMSI) in Bergen, Norway (1-4 May 1978), the following items were therefore discussed and partly decided upon (IOC, 1978) :

1. Organochlorine standards should be ; Lindane, DDD, DDE, DDT, DDMU and hexachlorobenzene. A separate standard was required for PCB, and Aroclor-1254 was chosen. The use of internal standards was discussed and it was concluded that they should be used.
2. Analytical procedures were discussed (Paris, 5-9 Semtember 1977) along the lines put forth by GEMSI at its First Session regarding the use of GCMS and gas chromatography with electron-capture detector. It was decided, however, to use gas chromatography unsupported by GCMS for the initial intercalibration phases of the baseline study. It was also decided not to limit the analytical procedure to one method. However, it was concluded that one preconcentration technique which is in use by several laboratories or was known, at that time, to produce acceptable results, could be presented in some detail. Other suitable isolation techniques needed only to be identified and listed, which was done (IOC, 1978 ; Annex IV thereof).

At the time of the Steering Committee meeting at the Bermuda Biological Station, 8-10 October 1979, the views on the analyses of PCB were somewhat changed because of the possibility of doing temperature-progammed, electron-capture, glass-capillary gas chromatography. A draft, « Outline of the method to be used for the determination of chlorinated hydrocarbons in sea water » was discussed and the method suggested was accepted. (IOC, 1980, Annex I thereof).

The activities under the organochlorine component of the intercalibration exercise consist of intercalibration of samples of open-ocean water to be collected at the Panulirus Station off Bermuda.

The analyses that were carried out during the intercalibration exercise at the Bermuda Biological Station were all conducted using temperature-programmed glass-capillary gas chromatography. As a result, it is now possible to report individual PCB components for the first time in an open-ocean water intercalibration study.

## II. 1. Equipment

(i) Water-column samplers

Several kinds of water sampler are available for the collection of water at depth.

### a) Bodman Bottle

The sea-water sampler suggested for collection of water volumes up to 90 litres is the Bodman sampler (Fig. II.1) This is a modification of the Bodman Sampler (Bodman *et. al.*, 1961) made by Bodega Marine Laboratory in collaboration with R. Hamblin, Oceanic Industries, Osterville, Mass. USA.

The main features of the Bodman sampler are :
1) construction with non-contaminating materials — anodized aluminium, stainless steel, teflon and viton ; 2) 45 kg disk lead weights bolted directly to the bottom of the sampler and a one way relief valve in the lid to permit subsurface cocking, such that the sampler could pass closed through the surface film, thereby minimizing a potential source of contamination ; 3) the cylinder, constructed of aluminium, was maintained at its full diameter at both ends, facilitating cleaning and enhancing flushing as it descends through the water column ; 4) the use of a Kullenberg piston-core release mechanism allows sampling at precise distances from the bottom ; 5) a magnetic switch on the upper plate of the sampler switches an integrally mounted 12 kHz pinger to double pulse rate at the time of the trip, thus permitting continuous shipboard monitoring of the position and condition of the sampler ; 6) a Swagelok quick-connect purge valve in the lid for the attachment of a nitrogen gas line, and a stainless steel ball valve with stan-

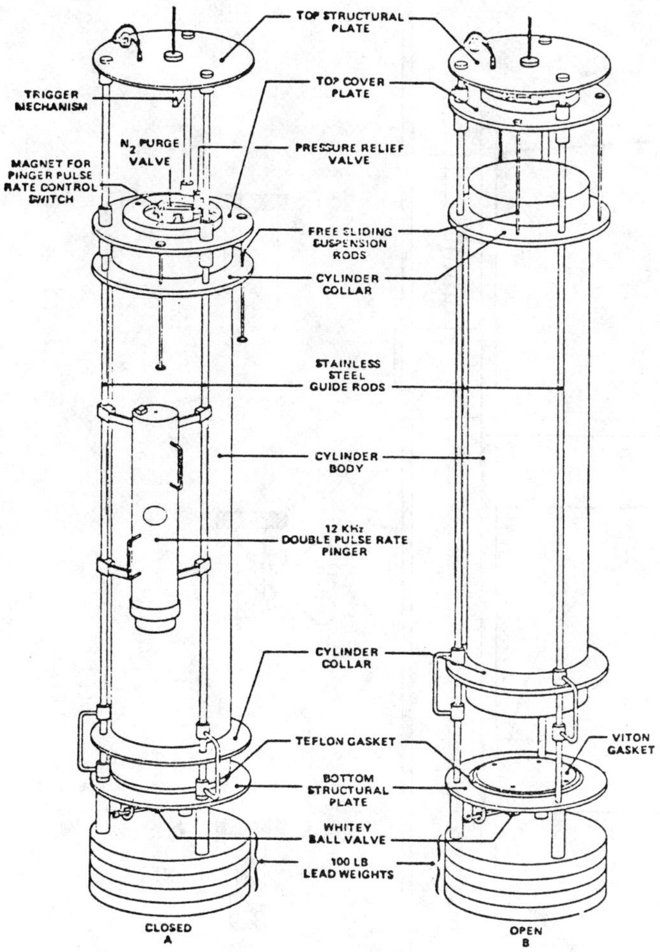


**Figure II.1 The Bodman sampler.**

dard Swagelok connections at the bottom of the sampler allows the contents of the Bodman bottle to be transferred directly to a cleanroom for precessing with limited exposure to the shipboard atmosphere ; 7) positioning of the outlet valve in the center of the flat bottom plate reduces the potential for loss of larger settling particulates between sample collection and processing, which has been reported as a deficiency in similar samplers.

Figure II.1 and the description of the sampler are copied from :
THE SAMPLING AND MEASUREMENT OF HYDROCARBONS IN NATURAL WATERS, by B.W. de Lappe, R.W. Risebrough, A.M. Springer, T.T. Schmidt, J.C. Shropshire, E.F. Letterman and J.R. Payne University of California, Bodega Marine Laboratory, Bodega Bay, California 94923 (unpublished paper).

Once obtained in the bottle the sample is pressurized with nitrogen and transferred under nitrogen into a precleaned stainless-steel drum (210 litres). This sampler is suitable for surface and deep waters and can be coupled by stainless-steel tubing to a sample reservoir. This prevents contamination by the ship's atmosphere.

### b) Glass-sphere water samplers

The German Hydrographic Institute uses à 10-litre glass-sphere sampler (Fig. II.2) that is attatched to the hydrographic wire, and the messenger in activated at the desired depth. At least three casts are required before a sample deemed large enough for analysis of organochlorines in open-ocean water in obtained. The flasks are only open to the atmosphere long enough to add solvent once that sample has been collected. The sampler has the advantage of being coupled with direct solvent extraction, the sample being extracted directly in the sampling vessel. This method was to be used as a comparison for the resin techniques.

Those samples collected by glass sphere water samplers were subjected to three extractions with aliquots of hexane on a reciprocating shaker. These were combined and analysed according to the recommented procedures.

### c) *In situ* resin column

The U.S. Texas A & M University laboratory brought an *in situ* XAD-2 resin column to Bermuda. This consists of a standard XAD-2 resin column attached to a stainless steel loop. This is suspended approximately ten metres below the water surface and a vacuum applied to obtain a flow rate of 250 ml $min^{-1}$.

### d) Gaslift system

A gaslift system (Tokar and Harvey, 1980) was used to obtain large quantities of water for analysis (Fig. II.3). This system employs 3/4'' stainless steel pipe connected by Swagelock quick connects. The pipe is attached to the hydrographic wire and lowered to approximately 10-m depth while adding 2-4 m pipe lengths. Nitrogen is injected at the base of the pipe through a perpendicular pipe and the rising and expanding nitrogen creates a vacuum. This results in a flow rate of about 8-10 l $min^{-1}$ at a depth of 10 metres. This system, when used in connection with stainless steel drums, also results in a sample free from contamination by the shipboard environment.

## (ii) Laboratory Facilities

Laboratories at the Bermuda Biological Station were especially prepared for the intercalibration exercise. With respect to the organochlorine component of the exercise, two laboratories were tiled after removal of all paint and possible sources of contamination. One laboratory was tiled in its entirety with ceramic tiles (i.e., ceiling, underside of benches etc.) This served as the organic clean-room, and was the site of the analytical procedures. This laboratory was equiped with a glass and aluminium fume hood, designed specifically for the exercise. A stainless-steel high-temperature (250° C) oven was provided for the baking of glassware. This oven was used only for this purpose and was dedicated to organochlorine analysis. All glassware was obtained from Ace Glass Co., Vineland, New Jersey. This included Kuderna-Danish evaporators, flasks, distillation apparatus, Soxhlet apparatus, heating mantles etc. All items were standard Ace Glass fittings obtainable through this company. Three Bucchi rotatory evaporators were also available for the concentration of samples.

Common sources of interference in the laboratory originate from glassware, chemicals and other materials used in the extraction and workup procedure, which have not been properly treated. Therefore, it is necessary to do routine and periodic checking of glassware and chemicals in order to ensure low blank values. For cleaning glassware the following procedure is recommented :

Wash all items in hot, soapy water and rinse thoroughly. Rinse each piece of glassware with pesticide-free acetone. Without drying the glassware, rinse each piece with pesticide-free hexane (Fig. II.4). Transfer the glassware to a drying oven and bake at 300°C for at least four hours, or at 250°C overnight.

A convenient procedure is to wash glassware by the described method each afternoon, place it in an oven equipped with an automatic timer set to shut off in the morning, so that it has time to cool by the time work commences each day. In this way, it can be used immediately and not exposed unnecessarily to the possibility of contamination.

The oven used in baking glassware and heating chemicals should be dedicated to chlorinated hydrocarbon analysis. Contamination, especially of chemicals such as $Na_2SO_4$, can take place if they are heated in an oven that has been used for other work.

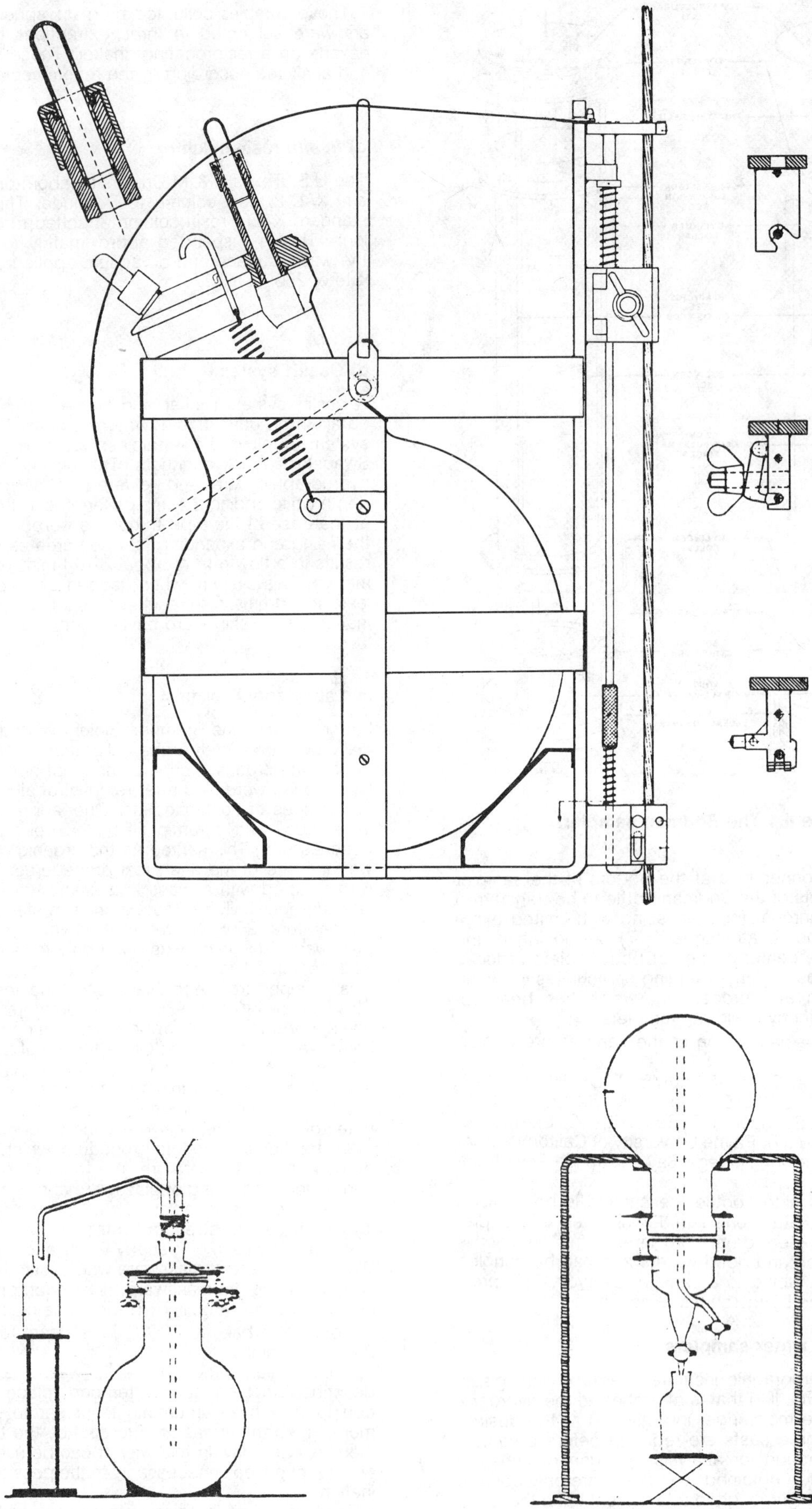

Figure II.2 Glass-sphere water sampler (10 litres in use at the German Huydrographic Institute (DHI).

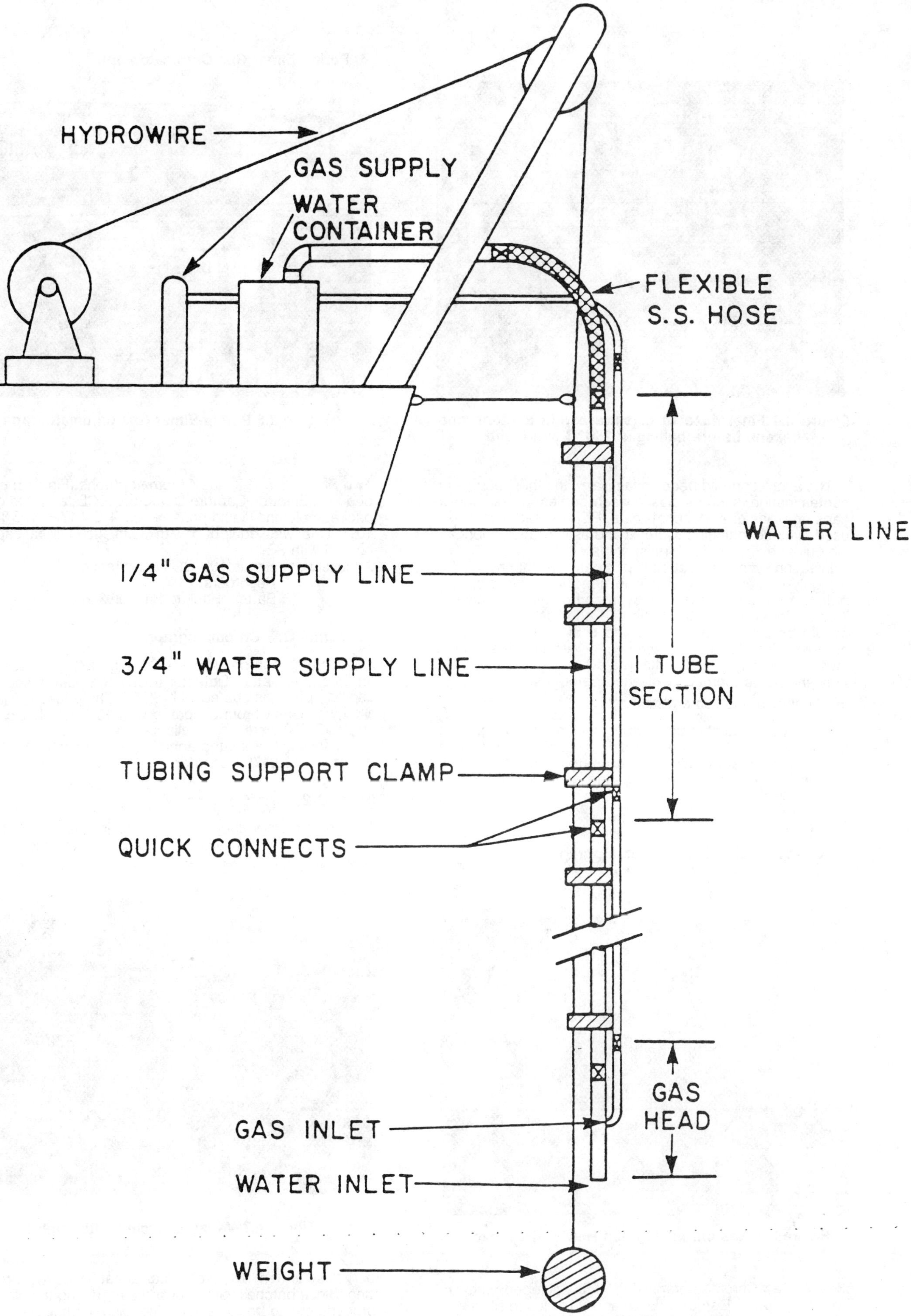


**Figure II.3 Diagram of the assembled gas lift rigged for continuous pumping. A 100 kg weight gives good stability and prevents bending of the tubes**

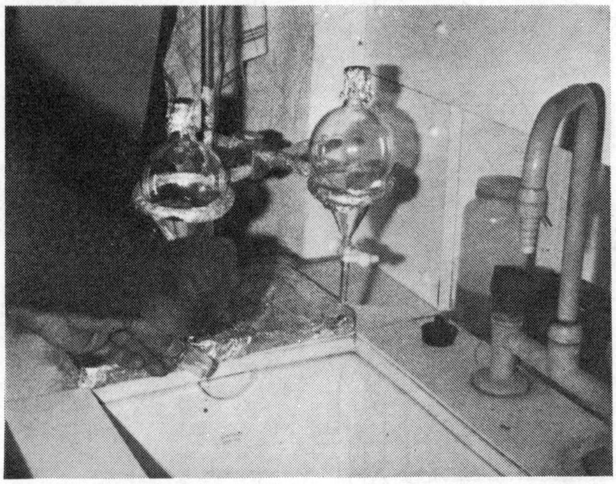

**Figure II.4 Final rinse of glassware with acetone and hexane before baking at 250°C overnight.**

The analyst should become familiar with other sources of contamination, such as plastic wash bottles, plastic screw caps on glass or metal containers, PCB-treated fibre glass used in air conditioning systems and as insulation, lubricating fluids, hydraulic fluid, paint chips etc.

Injection syringes must be kept clean. It is not unusual for a highly contaminated needle to need intensive cleaning before residual chlorinated hydrocarbons are removed. This could require 30-50 washings. Periodically, the syringe should be dismantled and soaked overnight in a suitable organic solvent such as acetone or ethylacetate. Since most new syringes are contaminated during manufacture, they should be thoroughly cleaned before use.

(iii) Analytical Instrumentation

The following gas chromatographs, with accessories, were made available from the respective manufacturers. The technical and application chemists from the respective companies were also present during an important part of the exercise.

**a) Hewlett-Packard Gas Chromatograph**

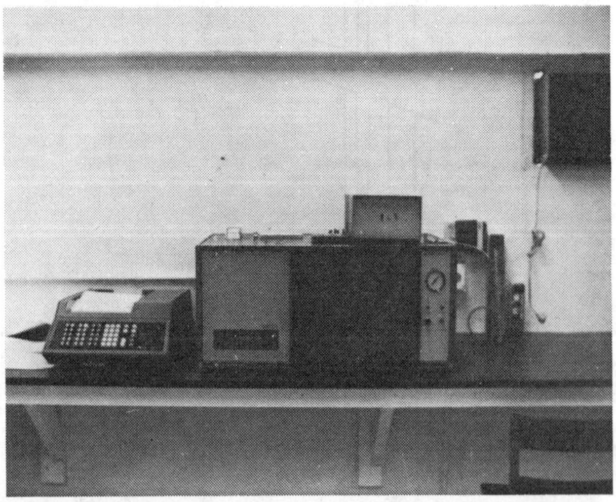

**Figure II.5 Gas chromatograph Hewlett-Packavol**

5840 A Gas Chromatograph (Fig. II.5) equipped with an Electron-Capture Detector (ECD) and 30-m fused-silica capillary columns coated with SE-54.
Accessories : 18835 B Capillary Inlet System
5840 A GC Terminal

**b) Perkin-Elmer Gas Chromatograph**

**Figure II.6 Perkin-Elmer Gas chromatograph**

Two Sigma 3 B GC Gas Chromatographs (Fig. ii.6) equipped with Electron-Capture Detectors (ECD) and 2-m glass-packed columns (liquid phase 1, 95 % OV-17 + 1, 5 % OV-101). One was equipped with 30-m fused-silica capillary coated with SE-54.
Accessories : Sigma 10B GC Data Station
Recorders
Battery Backup Pack 332-2400

**(c) Varian Gas Chromatograph**

Gas Chromatograph Model 3700 (Fig. ii.7) equipped with an Electron-Capture Detector (ECD) and 30-m fused-silica capillary columns coated with SE-54. The packed columns filled with Gas Chrom Q coated with 10 % OV 101 or with 10 % DC 200 were also available.
Accessories : Chromatographic Data Systems CDS 111
9176 Recorder

(iv) XAD-2 Resin Columns

The commercially supplied XAD-2 resin beads are usually contaminated with a variaty of organic compounds and fine

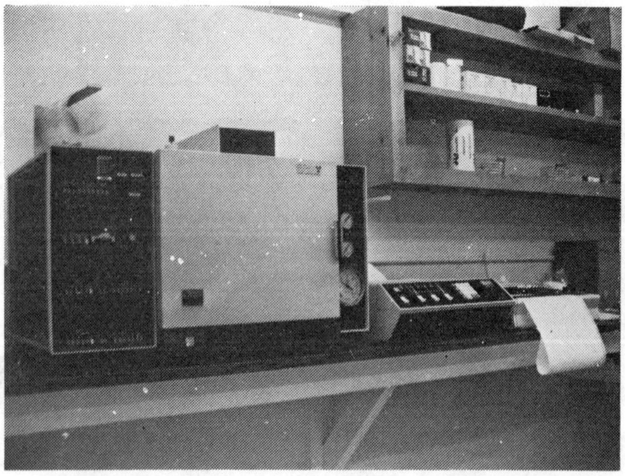

**Figure II.7 Varian Gas chromatograph**

particles formed by mechanical agitation during processing. Since batches vary, trial and error during the cleaning procedure is to be expected. The following has been fairly successful :

Shake the beads in a 50-60 mesh sieve to remove fine material and the wash those retained by the sieve with 30-

50 volumes of tap water. Extract the beads in a Sohxlet for at least 24 hours using acetonitrile as a solvent. Change the solvent and repeat the extraction for an additional 24 hours (Fig. II.8)

Sometimes at this stage the resin is sufficiently clean to use ; if not, additional extraction with different solvents, such as benzene or acetone, is necessary.

To check the progress of the clean-up procedure, a background blank determination is carried out.

To do this, pack a clean glass column (2 cm I.D.) with 50 cm³ of extracted resin. Elute the column by gravity flow with 200 ml of boiling acetonitrile. Dilute the eluate with 600 ml of pesticidefree water. Extract this aqueous acetonitrile solution with two 80-ml portions of hexane. Extract the combined hexane fractions with 10 ml of pesticide-free water. Dry the hexane extract over a minimum amount of anhydrous $Na_2SO_4$ and then concentrate it to 0.5 ml in a Kuderna-Danish concentrator or in a stream of dry nitrogen. Analyse that extract for interfering compounds on a gas chromatograph using the same procedure as will be employed for sea-water extracts. Calculate the interference on the basis of units of weight (usually ng) per litre. To do this, you must, of course, know at this point the amount of sea-water to be sampled.

If the blank is not acceptable, continue the cleaning process. When a satisfactory blank is obtained, the resin may be air dried and stored under pesticide-free water until needed.

**Fig. II.8 Reflux of XAD-2 resin with acetonitrile in a Sohxlet apparatus.**

(v) Chemicals, Standards etc.

**a) Solvents**

Solvents used were acetone, acetonitrile, hexane and methylene chloride. Two suppliers of solvents were used (Burdick and Jackson Laboratories, Inc., Muskegon, Michigan, U.S.A. and S.D.S. Solvents, Peypin, France). All solvents were pesticide grade, but blank determinations indicated that for the low concentrations of PCBs in open-ocean waters, further distillation was necessary.

**b) Other Chemicals**

Silica gel (60-200 mesh for chromatography), Florisil (100-200 mesh, chromatography grade) and sodium sulphate were purchased from Fischer Scientific Co., Pittsburg, Pennsylvania, USA.

Chemicals such as $Na_2SO_4$ and especially Florisil (R) have a tendency to adsorb contaminants from the atmosphere and should be periodically checked. $Na_2SO_4$ can be cleaned by thoroughly washing with benzene and methylene chloride, followed by heating to 300°C in a clean oven. For details about handling Florisil (R), the analyst should request the free bibliographies supplied by the manufacturer (Floridin, 3 Penn Center, Pittsburgh, Pa. 15235).

**c) Standards**

Standard Aroclor mixtures as well as individual components of 35 different isomers of polychlorinated biphenyls, having from 2 to 10 chlorine atoms, were provided by Analabs, Inc., North Haven, Connecticut, U.S.A.

Standards of selected organochlorine pesticides (DDT group) were also available. As internal standard the PCB isomer with four chlorine atoms not present in measurable amounts in the standard solutions was used.

## II.2 Outline of the analytical method for the determination of organochlorimes in sea water

### Introduction

At the Second Session of the GIPME Group of Experts on Methods, Standards and Intercalibration (GEMSI), in Bergen, Norway, 1-4 May 1978, a paper was produced on « ANALYSIS OF CHLORINATED HYDROCARBONS IN SEA-WATER BY ABSORPTION ON AMBERLITE XAD-2 RESIN ». This paper is used as the basis for the « outline of the method » described here, with a few modifications made by the Institute of Marine Research, Bergen, Norway and the International Laboratory of Marine Radioactivity, Monaco.

### Summary

In the method described here, seawater is passed through glass cartridges containing Amberlite XAD-2 resin which adsorbs the desired chlorinated hydrocarbons. The resin is then eluted with water-miscible organic solvent to remove the adsorbed compounds. The eluate is diluted with water and extracted with hexane. The hexane extract, containing chlorinated hydrocarbons is then subjected to solid-liquid column chromatography to remove unwanted substances, reduced in volume to give a convenient concentration, and finally analysed by gas chromatography. The overall efficiency of the method is about 90 % (± 10 %), for a number of chlorinated compounds including polychlorinated biphenyls and the DDT series.

The main advantage of the method is that after the adsorption step, the resin can be stored for many weeks without adversely affecting the results. So, the adsorption step can be effected with relative ease aboard ship, after which the cartridges are sealed and shipped to a land based laboratory for further analysis. Thus, the need for sophisticated laboratory facilities at sea is reduced.

In addition, since the adsorbtive capacity of the resin is quite high, the upper limit on sample size is imposed only by the ability to collect large volumes of seawater. In sampling surface waters, this difficulty can be overcome by drawing water directly from the ocean through a cartridge or resin.

### Sampling from ships

Most oceanographic vessels are floating contamination problems. Contaminating substances, especially PCBs, are potentially present in the ship's paint, hydraulic fluids, lubricating oils, anti-corrosives, hoses, bilge water, seawater plumbing systems etc. Every effort must be made to prevent seawater samples from coming in contact with these and other sources of contamination. It is essential to gain the co-operation of the officers and crew of the ship during a sampling programme in order to avoid such operations as bilge pumping or paint chipping when sampling gear is being used over the side.

Obviously, as with most analytical methods, certain steps require a lot of familiarity before they become routine, and probable sources of contamination must be minimized. Various aspects of the technique are presented in more detail below.

The analyses of seawater for organochlorines require collection of volumes in the order of 50-100 litres. The sampling can then either be done by collecting sea water by suction directly from the surface and upper layers of an appropriate volume through tygon or teflon tubing connected to a column with amberlite XAD-2 resin using a pump (Fig. II.9). The speed of the pump should be adjusted to deliver 250 ml min$^{-1}$. When samples below the surface are desired, the use of a Bodman Sampler, or Gas Lift system should be preferred (see Fig. II.1 and II.2).

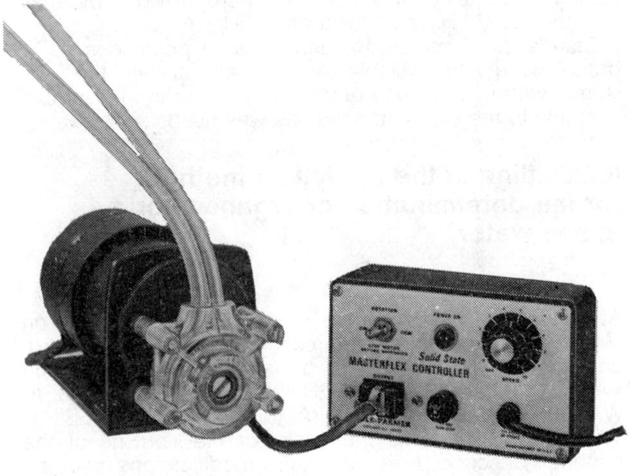


**Figure II.9 Cole Palmer pump used for extraction of seawater samples**

Adsorption of chlorinated hydrocarbons from seawater

Pack a 2 cm (i.d.) glass column with 50 ml of precleaned XAD-2 resin in organochlorine free water. Make sure that all air bubbles are removed. Pass the desired (measured) amount of seawater through the column at a rate of 200-250 ml min$^{-1}$. The adsorption efficiency of the resin is a function of the flow rate ; the rate given here is 4-5 bed volumes/min which has been experimentally determined to be the optimum and gives about 95 % adsorption efficiency for PCBs and DDTs (Harvey, Steinhauer and Teal, 1973). This can of course be accomplished by gravity flow or with a pump (Fig. II.10).

When using a pump it should be placed down stream of the column to avoid contamination.

**Figure II.10. Set-up for the adsorption of chlorinated hydrocarbons on XAD-2 resin in the Monaco laboratory**

It may be of interest to determine the relative amounts of chlorinated hydrocarbons in the dissolved and particulate states. However, filtering seawater before subjecting it to XAD-2 adsorption presents a problem because membrane, paper and glass-fibre filter materials adsorb organic matter.

Furthermore, membrane and paper filters usually contain substances which interfere with the analysis so that only precleaned glass-fibre filters are suitable. So, if one filters the sample of water and then analyzes the filter for particulate organochlorines the results tend to be high because the total amount of these compounds present is due to particulate content plus that adsorbed on the filter. By extension, if the filtrate is subjected to the resin adsorption technique, the « dissolved » concentrations will tend to be low because some of the organochlorines will have been lost on the filter. This problem can be overcome somewhat by first subjecting the sample to XAD-2 adsorption followed by filtration.

The reasons the latter is more accurate are twofold. First of all, since the resin beads are large compared to most seawater particulates, the particulates tend to pass through the resin bed and out of the column. Secondly, XAD-2 tends to pick up those organochlorines in the dissolved state rather than those adsorbed to particles. Only rarely does a sample of open-ocean seawater contain particles large enough to be held up by the column. These are readily visible at the top of the column and can be carefully removed before the column is eluted.

In the open ocean, the amount of organochlorines associated with particulates is usually less than 10 % of the total, whereas in coastal zones, it may account for 90 % or more. Therefore, the necessity of filtering seawater depends on the overall aims of the monitoring programme and the prevailing conditions at the sampling locations.

Internal standard

Before the elution of the XAD-2 column, an internal standard should be added to the top of the column if the behaviour is the same as that of the compounds investigated. The internal standard will then go through the entire work-up procedure and simplifies the quantification because there is no longer any need for exactly measured volumes and injections.

Extraction of resin and work-up of extract

The work-up of the resin can vary a little depending on the concentration of the organochlorines in the area to be monitored, the presence of interfering substances and the amount of water sampled. The following is suggested :

Elute the resin column with 200 ml of boiling acetonitrile under gravity flow (Fig. II.11).

The eluate is transferred to a separating funnel (specially made for bromine analysis) containing 600 ml pesticide-free water (Fig. II.12).

The acetonitrile/organochlorine-free water mixture is extracted twice with 80 ml portions of redistilled hexane.

Dry the extracts with a minimal amount of $Na_2SO_4$. After decanting the combined hexane fractions, wash the $Na_2SO_4$ with a few ml of fresh hexane and decant this into the combined hexane fraction. Carefully reduce the volume of the hexane solution to approx. 10 ml (do not exceed this limit or organochlorines can be lost !) on a rotary evaporator (Fig. II.13) ; a Kuderna-Danish evaporator might also be used (Fig. II.15).

The next step in the procedure is to remove interfering substances from the extracts — that is compounds which have GC retention-times that are the same or close to retention times of the target organochlorines. Quite often at this stage in the procedure the presence of interfering substances is exhibited by a yellow to brown tint in the extract. If not, then a preliminary check by electron-capture gas chromatography should be carried out.

If interfering substances are present they can usually be removed as follows :

Pass the concentrated extract through a micro-chromatography column (a Pasteur pipette will suffice) containing 2-4 cm of Florisil (R) (Fig. II.14). Wash the column with one

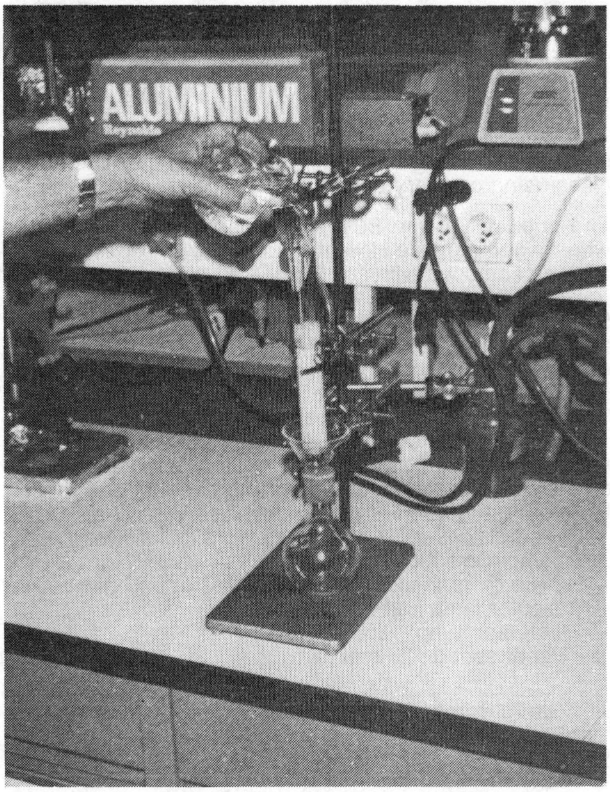


**Figure II.11. Elution of the adsorbed chlorinated hydrocarbons from the XAD-2 resin with boiling acetonitrile.**

bed volume of 5 % ether in hexane. The extract and ether-hexane washings are then combined and evaporated in a Kuderna-Danish concentrator to 0.5-1.0 ml (Fig. II.15) or in a stream of pure nitrogen.

It is important to quantify this volume for the final gas-chromatographic analyses if an internal standard is not used.

**Figure II. 12. Separating funnel for the hexane extraction of the acetonitrile eluate mixed with water.**

*(R) Registered trade mark*

After removing interfering substances, PCBs and p,p' -DDE can be separated from p,p' DDT and p,p' -DDD before electron-capture gaschromatographic analysis on packed columns as follows :

First, prepare deactivated silica by baking 100-200 mesh silica for 8 hrs at 140°C, then add to it 3 % (W/W) water. (Some experimentation with conditions using standard chlorinated hydrocarbon solutions as yields determinants is necessary). Pass the extracts (0.5-1.0 ml) through a microchromatography column packed with 2 grams of the deactivated silica. It is recommended that the microcolumns be pretreated with dichloromethane and subsequently with n-hexane in order to remove interfering substances (Duinker and Hillebrand, 1978). Elute the column with hexane (10-15 ml) which removes PCBs and p,p' -DDE, then elute with a 10 % ether-hexane solution (10-15 ml) or toluene which removes p,p'-DDT and p,p'-DDD. The eluates are then concentrated to a convenient volume for gas chromatographic analysis. Appropriate blanks should be run, repeating the analytical procedure using water that was extracted already. This gives an estimate of the presence of any contamination source, with the exception of the sampling procedure.

**Figure II. 13. Rotavapor for the reduction of the extraction volume.**

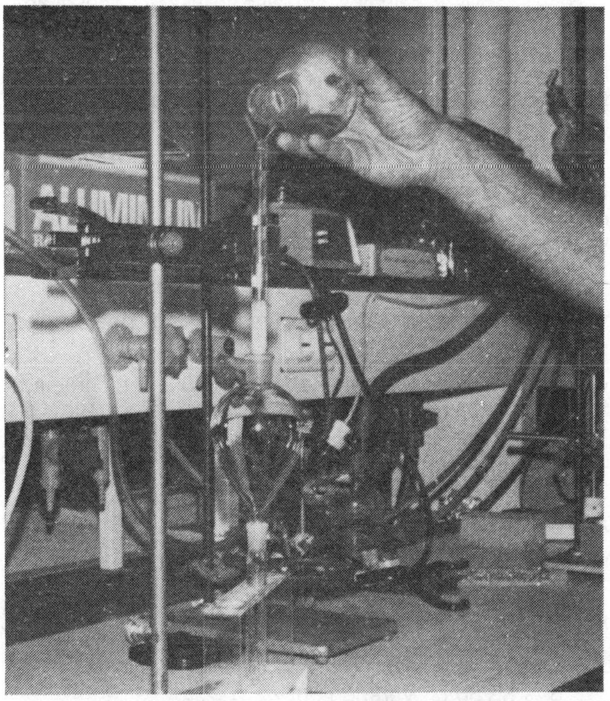

**Figure II.14. Short column with Florisil for the clean up of the extract (to remove, for example, phthalates and other organic compounds.**

**Figure II.15. Kuderna-Danish concentrator for final volume reduction.**

## Gaschromatographic analysis

There are several suitable gas-chromatographic stationary phases for organochlorine analysis. Information contained in the references at the end of this chapter may serve as a starting point for familiarization with materials commonly used.

Glass-capillary columns are also in use for these analyses and in that case the separation of PCBs and p,p'-DDE from the p,p' -DDT and p,p' -DDD is not necessary.

The gas chromatograph should be equipped with electron-capture detector. It should also have glass column(s) and all-glass injector and detector systems in order to avoid losses of samples by thermal « cracking ».

Sample extracts are injected into the GC using standard techniques. However, due to the small quantities of chlorinated hydrocarbons normally dealt with, special care should be taken.

### (i) Gas-chromatographic analyses using packed columns.

The analysis is performed on a Varian 3700. The chromatographic column used is a glass colmun filled with Gas Chrom Q coated with 10 % OV 101 or with 10 % DC 200.

For PCB analysis, different types of packing are used by laboratories which make routine analysis :

5 % DCFS 1265 + 4 % DC 200 on chromosorb W 80/100
3 % OV 210 + 3 % OV 17 on gas chrom Q 100/120
6 % QF1 on chromosorb W 80/100
10 % OV 17, 10 % OV 210 (1 : 4) on chromosorb W 80/100
1.5 % SP 2250 + 1.95 % SP 2401 on Supelcon aw
4 % SE 30 + 6 % OV 210 on gas chrom Q
5 % SE 30 on Varaport 30

The main problem is the homogeneity of the phase. Normally with 10 % (or more) of coating material there is no difficulty in obtaining an efficient column.

Experimental conditions :

| | |
|---|---|
| t° column : | 200°C |
| t° injector : | 210°C |
| t° detector : | 260°C |
| $N_2$ flow rate : | 50-70 ml min$^{-1}$ |
| Volume injected : | 1-10 µl |

### (ii) Gas-chromatographic temperature programmed analyses using capillary columns.

The analyses of the PCB and organochlorine pesticides were performed on a Hewlett Packard 5710 A gas chromatograph equipped with an HP 1874 A capillary-column control and a 45-m long glass-capillary column coated with SE-54. The construction of the electron-capture detector (ECD) permits the make-up gas to flow directly into the detector housing via a built-in metal coil which is heated by the detector and maintained at a constant temperature. Since the volume of the carrier gas entering the detector is very small compared to the make-up gas (~ d 1 :50), the temperature variation in the oven has little or no influence on the detector temperature and the capillary column can therefore be temperature programmed.

The conditions for the analyses were as follows :

Sample injection splitless,
Oven programmed from 100-230°C at 8°C min$^{-1}$
Injector temp. : 250°C
Detector temp. : 250°C
Paper speed : 20 mm min$^{-1}$

### (iii) Gas-chromatographic analysis and confirmatory tests

Before injecting the sample, each syringe should be checked for the amount of residual solvent contained in the needle portion. Generally, liquid contained in the needle at the completion of an injection is also evaporated onto the column, owing to the high temperature of the injector. In some cases, the amount of residual solvent can be in the range of 0.05-0.5 µl. This can change the calculated results considerably if not accounted for when injections in the µl range are being used or when an internal standard is not used.

Qualitative analysis of the chlorinated hydrocarbons in question is done by comparison of retention times for samples with standard solutions, preferably on at least two columns which differ considerably in the polarity of their stationary phases. Additional identification of the compounds of interest can be done by chemical alteration followed by qualitative analysis (as well as quantitative analysis in some cases). Such confirmation procedures are exemplified by the following procedure for the dehydrochlorination of p,p'-DDD and p,p'-DDT to form p,p'-DDMU (2-chloro-1, 1-bis (p-chlorophenyl)-ethylene and p,p'-DDE, respectively :

To each of two 0.1-0.5 ml aliquots of concentrated extract, contained in graduated centrifuge tubes, add 1 ml of 0.1 N $NaOCH_3$/methanol solution. Heat the two mixtures to 5O°C in a water bath. For DDT confirmation, heating should be for 30 mins ; for DDD, heat for 1 hr. Allow the two reaction mixtures to cool and then add 5 ml of pesticide-free water and 1 ml of pesticide-free hexane. Extract the newly formed derivatives by shaking the reaction mixtures vigorously ; then let stand until the hexane and aqueous phases separate. Remove the hexane phases. Re-extract the aqueous phases with an additional 1-ml portion of hexane. Combine the two extracts for each of the two experiments and reduce each in volume to a suitable concentration for gas-chromatographic analysis. DDE is oxidized by chromic acid.

Whenever possible, confirmation should be carried out using GC-MS techniques (mass-fragmentography).

For chemical derivation procedures to confirm the presence of other chlorinated hydrocarbon compounds the reader is referred to standard references such as « The Pesticide Analytical Manual » (Vol. I. U.S. Department of Health, Education and Welfare, Food and Drug Administration, Rockville, Md.)

Quantification of PCBs and DDT residues is accomplished by comparing peak heights in sample chromatograms with those produced by reference standards. The peaks in the chromatograms of sample extracts should be bracketed by those of standards because of the limited range of linear response by the ECD.

The following formula will give the concentration of a chlorinated hydrocarbon in a water sample :

$$[CH]\ ng\ l^{-1} = \frac{Htsam}{Htstd} \times \frac{WT}{Vsamp} \times \frac{Vext}{Vinj} \times 10^3$$

Where :

$[CH]$ ng $l^{-1}$ = concentration of chlorinated hydrocarbon in nanograms/litre.

$Ht_{sam}$ = peak height of chlorinated hydrocarbon on sample chromatogram.

$Ht_{std}$ = peak height of chlorinated hydrocarbon on standard solution chromatogram.

Wt = weight in nanograms of chlorinated hydrocarbon giving $Ht_{std}$.

$V_{samp}$ = volume in litres of water extracted.

$V_{ext}$ = final volume of extract in ml.

$V_{inj}$ = volume of extract in microlitres giving $Ht_{sam}$.

It should not be implied that the method outlined is the only method or necessarily the best method. It can, however, be considered as « state-of-the-art » along with other methods presented in recent literature references.

The following list of references includes those that describe techniques for isolating chlorinated hydrocarbons from seawater. In addition, there are some references which discuss problems with contamination ; a few references reporting results on samples using the techniques referenced and some containing descriptions of the advantages and disadvantages of the various techniques.

This list is by no means exhaustive, but contains the most important references describing the three main techniques commonly used for chlorinated hydrocarbon analysis.

### Example of GC-MS analyses of PCB a mixture of and organochlorine pesticide

A mixture of PCBs (Clophene-A50 equivalent of Aroclor 1254) and organochlorine pesticides was analysed on a computerized gas chromatograph mass spectrometer (Finnigan GC/MS/Ds Model 9000/3200F/6100) under the same chromatographic conditions and the same chromatographic colums as for the analyses performed using ECD-detector with temperature-programming (Fig. II. 16.) The degree of chlorination (Chlorine number) of the different PCB components was determined and this chromatogram was compared with the chromatogram obtained using the same column type, electron-capture detector and temperature programming. No attempt was made to determine the structure of the different isomers.

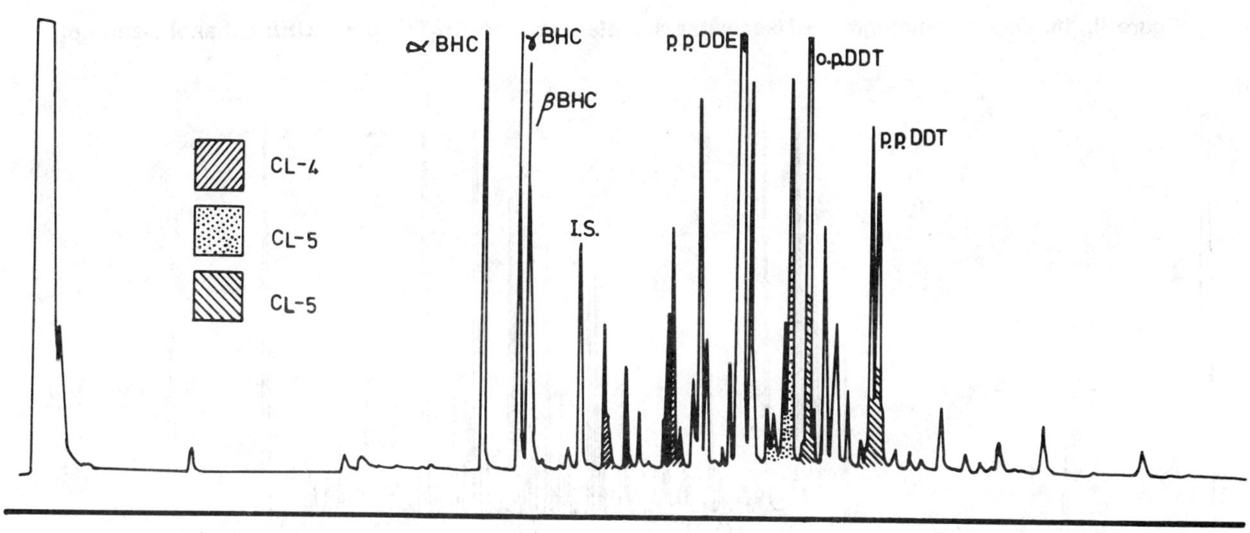


**Figure II. 16. Gas chromatogram of clophene-A50 and organochlorine pesticide mixture using EC-detector and temperature programming.**

**Figure II. 17. Gas chromatogram of sample after conc. $H_2SO_4$ clean-up.**

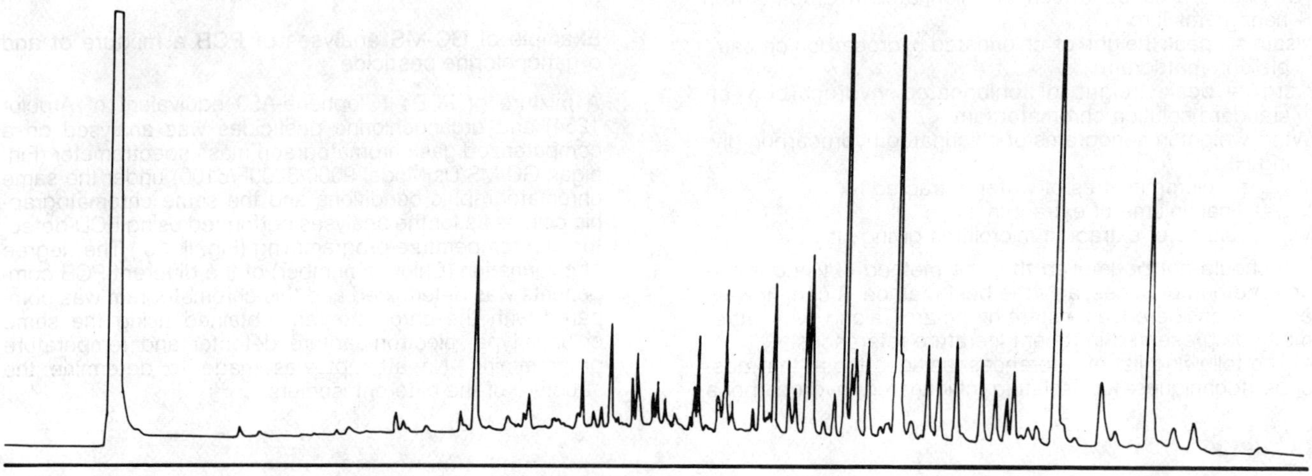

**Figure II. 18. Gas chromatogram of seawater sample after conc. $H_2SO_4$ and KOH/methanol clean-up.**

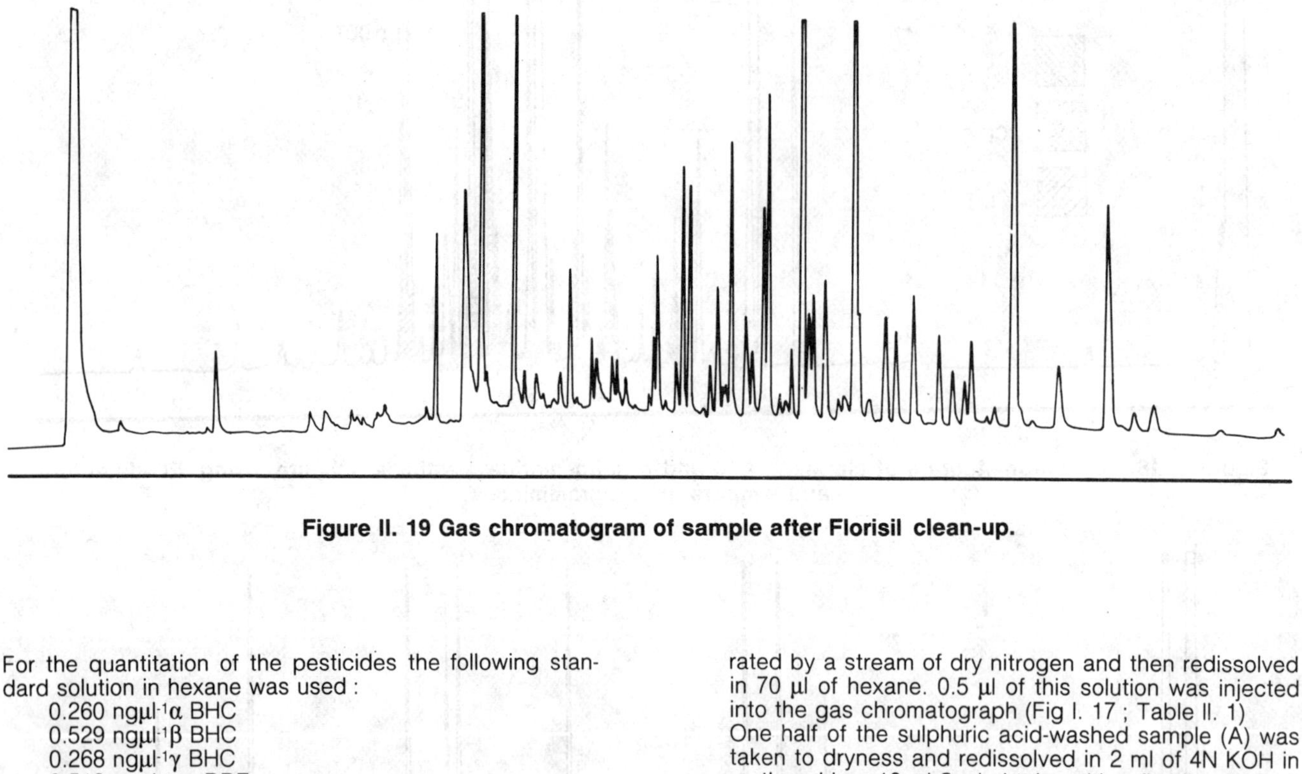

**Figure II. 19 Gas chromatogram of sample after Florisil clean-up.**

For the quantitation of the pesticides the following standard solution in hexane was used :

0.260 ngμl$^{-1}$α BHC
0.529 ngμl$^{-1}$β BHC
0.268 ngμl$^{-1}$γ BHC
0.510 ngμl$^{-1}$pp DDE
1.077 ngμl$^{-1}$op DDT
1.009 ngμl$^{-1}$pp DDT
0.600 ngμl$^{-1}$ internal standard (a PCB-$Cl_4$ isomer).

## Example of water sample analysis

The sample was subjected to the procedure described above and the $Na_2SO_4$-dried sample was reduced to 10 ml.

The 10-ml sample was divided into two parts A and B.

A. The 5-ml sample was mixed vigorously with conc. $H_2SO_4$ and centrifuged. The hexane-phase was evaporated by a stream of dry nitrogen and then redissolved in 70 μl of hexane. 0.5 μl of this solution was injected into the gas chromatograph (Fig I. 17 ; Table II. 1)
One half of the sulphuric acid-washed sample (A) was taken to dryness and redissolved in 2 ml of 4N KOH in methanol in a 10-ml Sovirel tube with teflon-lined screw cap. The tube was left at 80°C, for one hour. The sample was diluted with approx. 0.5 ml distilled water and extracted with 2 × 3 ml of hexane. The hexane-phase was evaporated to dryness and redissolved in 50 μl of hexane and 0.7 μl was injected into the gas chromatograph (Fig II. 18 ; Table II. 2)

B. The volume of 5 ml was adjusted to 10 ml and subjected to clean-up on a Florisil column (~ 0.5 × 4 cm). The eluate was evaporated to dryness, redissolved in 100 μl of hexane and 0.5 μl was injected into the gas chromatograph (Fig II. 19 ; Table II. 3)

**Table II.1.**

*Content of PCB and pesticides in nanogram per litre sea water, $H_2SO_4$ clean-up. (Perkin Elmer PEP-1 data system.)*

| TIME | ARFA | RRT | RF | C | NAME |
|---|---|---|---|---|---|
| 14.59 | .0683 | .810, | .0000, | .0000, | |
| 14.75 | .0446 | .818, | .0000, | .0000, | |
| 14.80 | .0448 | .821, | .0000, | .0000, | |
| 14.98 | .3721 | .831, | .0000, | .0000, | |
| 15.11 | .3069 | .838, | .0000, | .0000, | |
| 15.32 | 26.8889 | .850, | .3162, | 1.5096, | ALFA-BHC : |
| | | .857, | .0955, | | HEXACLOROBENZENE : |
| 15.55 | 1.1011 | .863, | .0000, | .0000, | |
| 15.71 | .7341 | .872, | .0000, | .0000, | |
| 16.00 | .1058 | .888, | .0000, | .0000, | |
| 16.07 | .0715 | .892, | .0000, | .0000, | |
| 16.16 | 1.1744 | .897, | .0000, | .0000, | |
| 16.30 | 24.4416 | .905, | .3152, | 1.3672, | GAMMA-BHC : |
| 16.61 | 2.1062 | .922, | .3995, | .1494, | BETA-BHC : |
| 16.80 | .9382 | .932, | .0000, | .0000, | |
| 17.04 | 1.1864 | .946, | .0000. | .0000, | |
| 17.16 | 4.4830 | .952, | .0000, | .0000, | |
| 17.41 | .2967 | .966, | .0000, | .0000, | |
| 17.51 | 3.5308 | .972, | .0000, | .0000, | |
| 17.72 | 1.0532 | .983, | .0000, | .0000, | |
| 17.84 | .0689 | .990, | .0000, | .0000, | |
| 18.01 | 4.5049 | 1.000, | 1.0000, | .8000, | INTERNAL STANDARD : |
| 18.07 | 2.5641 | 1.003, | 1.0000, | .4553, | |
| 18,21 | .3009 | 1.011, | .0000, | .0000, | |
| 18.33 | .1569 | 1.017, | .0000, | .0000, | |
| 18.49 | .1010 | 1.026, | .0000, | .0000, | |
| 18.65 | 1.2641 | 1.035, | 1.8491, | .4150, | 7 PCB-CL4 : |
| 18.79 | 2.6540 | 1.043, | .9413, | .4436, | 8 PCB-CL4 : |
| 19.09 | .0837 | 1.059, | .0000, | .0000, | |
| 19.27 | .9950 | 1.069, | .9900. | 749, | 11 PCB-CL4 : |
| 19.40 | 1.5246 | 1.077, | .0000, | .0000, | |
| 19.52 | .2719 | 1.083, | .0000, | .0000, | |
| 19.64 | 1.2592 | 1.090, | .4347, | .0972, | 13 PCB-CL4 : |
| 19.91 | .3143 | 1.105, | .0000, | .0000, | |
| 20.09 | .1520 | 1.115, | .0000, | .0000, | |
| 20.34 | .5323 | 1.129, | .4754, | .0449, | 15 PCB-CL4 : |
| 20.48 | 1.7389 | 1.137, | .5679, | .1753, | 16 PCB-CL4 : |
| 20.59 | 3.0278 | 1.143, | .6325, | .3400, | 17 PCB-CL5 : |
| 20.82 | .6977 | 1.156, | .4629, | .0573, | 18 PCB-CL5 : |
| 20.99 | .6874 | 1.165, | .0000, | .0000, | |
| 21.14 | 1.9122 | 1.173. | .2595, | .0881, | 19 PCB-CL5 : |
| 21.34 | 4.1318 | 1.184, | .3316, | 2432, | 21 PCB-CL5 : |
| 21.55 | 19.5803 | 1.196, | .3731, | 1.2950, | 22 PCB-CL5 |
| 21.96 | .1686 | 1.219, | .3426, | .0102, | 23 PCB-CL5 : |
| 22.16 | .6867 | 1.230, | .3623, | .0441, | 24 PCB-CL5 : |
| 22.37 | 2.5510 | 1.242, | .0969, | .0438, | PP-DDE : |
| 22.54 | .5349 | 1.251, | .1245, | .0118, | DIELDRIN : |
| 22.62 | .7224 | 1.255, | .1245, | .0159, | DIELDRIN : |
| 22.76 | 4.7523 | 1.263, | .2318, | .1956, | 29 PCB-CL5 : |
| 23.20 | 2.3484 | 1.288, | 1.1964, | .0777, | 30 PCB-CL6 : |
| 23.39 | 1.8882 | 1.298, | .1063, | .0356, | 31 PCB-CL6 : |
| 23.73 | 4.2638 | 1.317, | .2238, | .1694, | 33 PCB-CL6 : |
| 23.87 | 5.0465 | 1.325, | .1210, | .1084, | 34 PCB-CL5 : |
| 24.09 | .6238 | 1.337, | .0000, | .0000, | |
| 24.23 | .5931 | 1.345, | .0000, | .0000, | |
| 24.40 | .6863 | 1.354, | .3710, | .0452, | PP-DDD-OP-DDT : |
| 24.56 | 1.2054 | 1.363, | .2305, | .0493, | 36 PCB-CL6 : |
| 24.86 | 16.3014 | 1.380, | .0989, | .2860, | 37 PCB-CL6 : |
| 25.06 | 2.3361 | 1.391, | .1976, | .0819, | 38 PCB-CL6 : |
| 25.24 | 15.0214 | 1.401, | .1022, | .2720, | PCB-CL5 : |
| 25.53 | 3.9404 | 1.417, | .0710, | .0496, | 40 PCB-CL6 : |
| 25.78 | .0866 | 1.431, | .00000, | .0000, | |
| 25.91 | .5202 | 1.438, | .0000, | .0000, | |
| 26.07 | .4403 | 1.447, | .0000, | .0000, | |
| 26.18 | .4781 | 1.453, | .5493, | .0466, | PP-DDT : |
| 26.38 | 17.2108 | 1.464, | .0691, | .2108, | 42 PCB-CL6 : |
| 26.51 | 2.6006 | 1.471, | .0691, | .0318, | 42 PCB-CL6 : |
| 26.80 | 1.0094 | 1.488, | .0000, | .0000, | |
| 27.13 | .2284 | 1.506, | .0000, | .0000, | |
| 27.11 | 3.3273 | 1.516, | .0000, | .0000, | |
| 27.62 | 3.0499 | 1.533, | .0000, | .0000, | |
| 28.15 | 3.4651 | 1.563, | .0529, | .0324, | 43 PCB-CL6 : |
| 28.34 | .4352 | 1.573, | .0000, | .0000, | |
| 28.90 | 3.1355 | 1.604, | .0000, | .0000, | |
| 29.31 | 2.1676 | 1.627, | .0000, | .0000, | |
| 29.64 | 1.5803 | 1.645, | .0000, | .0000, | |
| 29.84 | 2.6908 | 1.656, | .0000, | .0000, | |

## Table II.2.

*Content of PCB and pesticides in nanogram per litre sea water, $H_2SO_4$ and KOH/meth clean-up.*

| TIME | ARFA | RRT | RF | C | NAME |
|---|---|---|---|---|---|
| 14.75 | .0548 | .819, | .0000, | .0000, | |
| 14.96 | .8621 | .831, | .0000, | .0000, | |
| 15.15 | .1983 | .841, | .0000, | .0000, | |
| 15.26 | .1485 | .847, | .3162, | .0125, | ALFA-BHC : |
| 15.33 | .0684 | .851, | .3162, | .0057 | ALFA-BHC : |
| 15.43 | .4556 | .857, | .0955, | .0116, | HEXACLOROBENZENE : |
| 15.56 | 1.1134 | .864, | .0000, | .0000, | |
| 15.81 | .1747 | .878, | .0000, | .0000, | |
| 16.08 | .1541 | .893, | .0000, | .0000, | |
| 16.35 | .6093 | .908, | .3152, | .0000, | GAMA-BHC : |
| 16.45 | .4569 | .913, | .0000, | .0000, | |
| 16.61 | .2130 | .922, | .3995, | .0227, | BETA-BHC |
| 16.70 | .1226 | .927, | .0000, | .0000, | |
| 16.79 | .1141 | .932, | .0000, | .0000, | |
| 17.05 | .7395 | .947, | .0000, | .0000, | |
| 17.17 | 1.7264 | .953, | .0000, | .0000, | |
| 17.38 | .1486 | .965, | .0000, | .0000, | |
| 17.51 | .5409 | .972, | .0000, | .0000, | |
| 17.74 | .7255 | .985, | .0000, | .0000, | |
| 18.00 | 2.9976 | 1.000, | 1.0000, | .8000, | INTERNAL STANDARD : |
| 18.13 | .2099 | 1.007, | .0000, | .0000, | |
| 18.23 | .2823 | 1.012, | .0000, | .0000, | |
| 18.33 | .1650 | 1.018, | .0000, | .0000, | |
| 18.50 | .1842 | 1.027, | .0000, | .0000, | |
| 18.66 | 1.1192 | 1.036, | 1.8491, | .5523, | 7 PCB-CL4 : |
| 18.90 | 2.7265 | 1.044, | .9413, | .6848, | 8 PCB-CL4 : |
| 19.03 | .1143 | 1.057, | .0000, | .0000, | |
| 19.13 | .1259 | 1.062, | .0000, | .0000, | |
| 19.28 | .9675 | 1.071, | .9900, | .2556, | 11 PCB-CL4 : |
| 19.41 | 1.3973 | 1.078, | .0000, | .0000, | |
| 19.65 | 1.0380 | 1.091, | .4347, | .1204, | 13 PCB-CL4 : |
| 19.93 | .5236 | 1.107, | .0000, | .0000, | |
| 20.10 | .2198 | 1.116, | .0000, | .0000, | |
| 20.24 | .0338 | 1.124, | .0000, | .0000, | |
| 20.36 | .3422 | 1.131, | .4754, | .0434, | 15 PCB-CL4 : |
| 20.48 | 1.0559 | 1.137, | .5679, | .1600, | 16 PCB-CL4 : |
| 20.60 | 2.4882 | 1.144, | .6325, | .4200, | 17 PCB-CL5 : |
| 20.81 | .3714 | 1.156, | .4629, | .0459, | 18 PCB-CL5 : |
| 21.16 | 1.6808 | 1.175, | .2595, | .1164, | 19 PCB-CL5 : |
| 21.35 | 4.3332 | 1.186, | .3316, | .3834, | 21 PCB-CL5 : |
| 21.54 | .9524 | 1.196, | .3731, | .0948, | 22 PCB-CL5 : |
| 21.77 | .3254 | 1.209, | .0000, | .0000, | |
| 21.97 | .1376 | 1.220, | .3426, | .0125, | 23 PCB-CL5 : |
| 22.17 | .7241 | 1.231, | .3623, | .0700, | 24 PCB-CL5 : |
| 22.41 | 3.3579 | 1.245, | .0969, | .0868, | PP-DDE : |
| 22.55 | .4544 | 1.252, | .1245, | .0150, | DIELDRIN : |
| 22.63 | .6233 | 1.257, | .1245, | .0207, | DIELDRIN : |
| 22.77 | 4.3872 | 1.265, | .2318, | .2713, | 29 PCB-CL5 : |
| 23.20 | 2.3783 | 1.288, | .1864, | .1182, | 30 PCB-CL6 : |
| 23.40 | 1.6402 | 1.300, | .1063, | .0465, | 31 PCB-CL6 : |
| 23.74 | 4.2664 | 1.318, | .2238, | .2548, | 33 PCB-CL6 : |
| 23.87 | 5.5763 | 1.326, | .1210, | .1800, | 34 PCB-CL5 : |
| 24.24 | .5384 | 1.346, | .0000, | .0000, | |
| 24.42 | .1185 | 1.356, | 3710, | .0117, | PP-DDD-OP-DDT : |
| 24.47 | 1.3135 | 1.365, | .2305, | .0809, | 36 PCB-CL6 : |
| 24.87 | 16.2931 | 1.381, | .0989, | .4296, | 37 PCB-CL6 : |
| 25.07 | 2.3685 | 1.392, | .1976, | .1248, | 38 PCB-CL6 : |
| 25.19 | 3.4953 | 1.399, | .1022, | .0952, | 39 PCB-CL5 : |
| 25.54 | 3.6960 | 1.418, | .0710, | .0700, | 40 PCB-CL6 : |
| 25.78 | .0623 | 1.432, | .0000, | .0000, | |
| 25.92 | .4630 | 1.440, | .0000, | .0000, | |
| 26.15 | .9818 | 1.452, | .5493, | .1439, | PP-DDT : |
| 26.39 | 18.1254 | 1.466, | .0691, | .3340, | 42 PCB-CL6 : |
| 26.52 | .25748 | 1.473, | .0691, | .0474, | 42 PCB-CL6 : |
| 26.81 | .10300 | 1.489, | .0000, | .0000, | |
| 27.13 | .1765 | 1.507, | .0000, | .0000, | |
| 27.31 | 3.5110 | 1.517, | .0000, | .0000, | |
| 27.63 | 3.0632 | 1.535, | .0000, | .0000, | |
| 27.94 | .0458 | 1.552, | .0000, | .0000, | |
| 28.15 | 3.8857 | 1.561, | .0529, | .0548, | 43 PCB-CL6 : |
| 28.34 | .4374 | 1.574, | .0000, | .0000, | |
| 28.90 | 3.2577 | 1.605, | .0000, | .0000, | |
| 29.15 | .0535 | 1.619, | .0000, | .0000, | |
| 29.31 | 2.1982 | 1.628, | .0000, | .0000, | |
| 29.65 | 1.6590 | 1.647, | .0000, | .0000, | |
| 29.85 | 2.8483 | 1.658, | .0000, | .0000, | |

**Table III.3.**

*Content of PCB and pesticides in nanogram litre sea water, Florisil clean-up.*

| TIME | AREA | RRT | RF | C | NAME |
|---|---|---|---|---|---|
| 14.61 | .0684 | .811, | .0000, | .0000, | |
| 14.84 | 15.6908 | .824, | .0000, | .0000, | |
| 15.32 | 23.4137 | .851, | .3162, | 1.5104, | ALFA-BHC : |
| | | .857, | .0955, | | HEXACLOROBENZENE : |
| 15.55 | 2.4216 | .863, | .0000, | .0000, | |
| 15.80 | .8506 | .877, | .0000, | .0000, | |
| 16.07 | .4448 | .892, | .0000, | .0000, | |
| 16.31 | 11.3100 | .906, | .3152, | .7276, | GAMMA-BHC : |
| 16.44 | .9587 | .913, | .0000, | .0000, | |
| 16.64 | 1.1873 | .924, | .3995, | .0968, | BETA-BHC : |
| 17.00 | 1.4422 | .944, | .0000, | .0000, | |
| 17.20 | .5449 | .955, | .0000, | .0000, | |
| 17.37 | .1045 | .965, | .0000, | .0000, | |
| 17.51 | .4627 | .972, | .0000, | .0000, | |
| 17.73 | 1.3096 | .985, | .0000, | .0000, | |
| 18.00 | 3.9190 | 1.000, | 1.0000, | .0000, | INTERNAL STANDARD : |
| 18.22 | .3342 | 1.012, | .0000, | .0000, | |
| 18.33 | .1041 | 1.018, | .0000, | .0000, | |
| 18.51 | .1337 | 1.028, | .0000, | .0000, | |
| 18.66 | 1.7055 | 1.036, | 1.8491, | .6437, | 7 PCB-CL4 : |
| 18.80 | 3.1187 | 1.044, | .9413, | .5992, | 8 PCB-CL4 : |
| 19.27 | 1.6357 | 1.070, | .9900, | .3305, | 11 PCB-CL4 : |
| 19.40 | 1.5786 | 1.077, | .0000, | .0000, | |
| 19.64 | 1.3436 | 1.091, | .4347, | .1192, | 13 PCB-CL4 : |
| 19.92 | .2213 | 1.106, | .0000, | .0000, | |
| 20.14 | .0223 | 1.118, | .0000, | .0000, | |
| 20.23 | .0632 | 1.123, | .0000, | .0000, | |
| 20.35 | .4706 | 1.130, | .4754, | .0456, | 15 PCB-CL4 : |
| 20.48 | 1.8369 | 1.137, | .5679, | .2129, | 16 PCB-CL4 : |
| 20.59 | .41118 | 1.143, | .6325, | .5308, | 17 PCB-CL5 : |
| 20.81 | .5360 | 1.156, | .4629, | .0506, | 18 PCB-CL5 : |
| 21.15 | 2.0829 | 1.175, | .2595, | .1103, | 19 PCB-CL5 : |
| 21.34 | 6.5504 | 1.185, | .3315, | .4432, | 21 PCB-CL5 : |
| 21.55 | 6.4227 | 1.197, | .3731, | .4890, | 22 PCB-CL5 : |
| 21.76 | .3553 | 1.208, | .0000, | .0000, | |
| 21.97 | .3443 | 1.220, | .3426, | .0240, | 23 PCB-CL5 : |
| 22.17 | 1.5580 | 1.231, | .3623, | .1152, | 24 PCB-CL5 : |
| 22.37 | 4.1049 | 1.242 | .0969, | .0811, | PP-DDE : |
| 22.55 | .7885 | 1.252, | .1245, | .0200, | DIELDRIN : |
| 22.63 | .9328 | 1.257, | 1245, | .0237, | DIELDRIN : |
| 22.76 | 8.0873 | 1.264, | 2318, | .3524, | 29 PCB-CL5 : |
| 23.20 | 3.3360 | 1.288, | 1864, | .1266, | 30 PCB-CL6 : |
| 23.40 | 2.4016 | 1.300, | 1063, | .0520, | 31 PCB-CL6 : |
| 23.73 | 6.1632 | 1.318, | 2238, | .2614, | 33 PCB-CL6 : |
| 23.87 | 9.6313 | 1.326, | 1210, | .2376, | 34 PCB-CL5 : |
| 24.24 | .6689 | 1.346, | .0000, | .0000, | |
| 24.40 | .5670 | 1.355, | 3710, | .0429, | PP-DDD-OP-DDT : |
| 24.57 | 2.0590 | 1.365, | .2305, | .0968, | 36 PCB-CL6 : |
| 24.86 | 20.6572 | 1.381, | .0989, | .4168, | 37 PCB-CL6 : |
| 25.07 | 3.4169 | 1.392, | 1976, | .1377, | 38 PCB-CL6 : |
| 25.20 | 4.5241 | 1.400, | 1022, | .0943, | 39 PCB-CL5 : |
| 25.54 | 5.0256 | 1.418, | 0710, | .0728, | 40 PCB-CL6 : |
| 25.92 | .8550 | 1.440, | .0000, | .0000, | |
| 26.09 | 1.5120 | 1.449, | .0000, | .0000, | |
| | | 1.457, | .5493 | | PP-DDT : |
| 26.39 | 23.2243 | 1.466, | .0691, | .3264, | 42 PCB-CL6 : |
| 26.51 | 3.3347 | 1.472, | .0691, | .0469, | 42 PCB-CL6 : |
| 26.85 | 1.4712 | 1.491, | .0000, | .0000, | |
| 27.12 | .1536 | 1.506, | .0000, | .0000, | |
| 27.31 | 3.8182 | 1.517, | .0000, | .0000, | |
| 27.63 | 3.4809 | 1.535, | .0000, | .0000, | |
| 28.16 | 4.7886 | 1.564, | .0529, | .0516, | 43 PCB-CL6 : |
| 28.34 | .4504 | 1.574, | .0000, | .0000, | |
| 28.63 | .1949 | 1.590, | .0000, | .0000, | |
| 28.91 | 3.6080 | 1.606, | .0000, | .0000, | |
| 29.32 | 2.2690 | 1.628, | .0000, | .0000, | |
| 29.66 | 1.6956 | 1.647, | .0000, | .0000, | |
| 29.85 | 3.3435 | 1.658, | .0000, | .0000, | |

**Figure II.20 The Bodmann Bottle is prepared for sampling of sea water for organochlorine analysis**

## II.3 Laboratory and shipboard activities at Bermuda

### Preparatory work

Some participants who arrived at the Bermuda Biological Station before the start of shipboard operations undertook preparatory work, such as checking the gas-chromatographic procedures and instruments, checking and controlling blanks, and the preparation of sampling equipment. All glassware was rinsed in water and soap, then rinsed with acetone and hexane and heated at 250°C overnight. The glassware was stored capped with aluminium foil that had been rinsed with n-hexane. Carboys and drums obtained in Bermuda were also rinsed with acetone and hexane. Florisil and sodium sulphate were Soxhlet-extracted with n-hexane for 12 hours.

A first meeting of the participants in the organochlorine group was held on 11 January 1980. The scientists taking part in the exercise were identified and an operational plan for the first sampling cruise was developed.

### The first cruise

The objective of the first cruise was to determine the concentration of chlorinated hydrocarbons at 1200- and 10-metre depths, in order to estimate the size of the sample required for the intercalibration experiment. Rough weather during the cruise resulted in the loss of one large-volume (90-l) ; Bodman bottle, after obtaining only one sample at 10-m depth. Sampling with this type of device was thus abandoned because of the danger involved in its use. However, sub-surface water (10 m) was sampled, using the Tokar and Harvey gas-lift system, which resulted in the collection of 420 litres of sea water. The seawater was stored in two 210-litre stainless-steel drums which had been previously only for the storage of sea water.

Sea water from the drums was immediately passed through glass columns containing XAD-2 resin (five bed volumes per minute : 250 ml $min^{-1}$), resulting in two extracts of 50 litres each and one of 126 litres. These samples were transferred to the laboratory and analysed according to the described procedure. The resulting chromatograms appeared to be rather complicated, and their interpretation was hampered by the large number of peaks present. It was suspected that some of these peaks might be contamination. Blanks were run through the entire procedure, using one XAD-2 column five times in succession. Resulting chromatograms showed relatively high blank values, especially before florisil treatment.

The extract of each sample was analysed by a gas chromatograph equipped with and electron-capture detector (ECD) ; samples obtained in the Institute of Marine Research in Bergen and in the International Laboratory of Marine Radioactivity in Monaco (both from 10-m depth), obtained under similar chromatographic conditions to those obtained in the Bergen Laboratory, were available for comparison.

Essentially, no individual components of polychlorinated biphenlys were used at that stage. Semiquantitative estimation of PCB in the same way as had been reported before, based on comparison of packed-column chromatograms of sample and some selected technical formulation, resulted in 0.05 (ng $l^{-1}$ PCB) (Aroclor 1254 equivalent). This was an order of magnitude lower than any value reported before, based on unpublished (Harvey, 1980, and Elder and Villeneuve, 1980 pers. comm.) and published (Duinker and Hillebrand, 1979) data.

It has been recognized that the retention times of dominant peaks in the sample chromatogram were shorter than those of Aroclor 1254 peaks. Duinker and Hillebrand (1979) found that less highly chlorinated components represent the PCB composition in North Sea water as well as or even better than the commonly accepted Aroclor 1254.

It appeared that several peaks in the sample chromatogram had retention times identical to peaks typical of Aroclor 1221 and 1242. Therefore, it was decided to compare the chromatograms of the samples with chromatograms of as many individual PCB components as were obtainable during the Workshop. For this purpose standard solutions of sixteen mono, di-, tri-, and tetra- chlorobiphenyls were prepared at various concentrations. These, and other standards still to be prepared, were to be used in identifying the peaks present in the chromatograms obtained in the analyses of sea-water samples.

### The secund cruise

The preparations for the second cruise were made according to the requirements set forth in the operational plan for sea-water sampling for organochlorine analysis, and further developed as a result of knowledge gained from the analyses of samples collected on the first cruise.

In order to improve the quality of the blanks of the XAD-2 resin to be used on the second cruise to obtain samples for the workshop and samples to be taken back to the different laboratories, XAD-2 was refluxed with acetonitrile for 24 hours, and packed using the slurry method.

The second cruise had the following objectives

(i) To compare three different methods of sample collection and storage
(a) a glass-sphere water sampler (German Hydrographic Institute) with solvent extraction in the sampler itself ;
(b) collection of water in steel drums, filled with water taken with a 90-litre Bodman sampler, followed by XAD-2 resin column extraction ;
(c) collection of water in steel drums through the gaslift system followed by the XAD-2 resin column extraction.
(ii) To fill a 1000-litre steel container and simultaneously pass water from it through 14 XAD-2 resin columns, to allow each participant to analyse one sample at the Bermuda Biological Station for Research, Inc., and one sample later in his home laboratory.
(iii) To fill glass and steel containers with 700 litres of water for multiple analyses by all participants in the Bermuda Biological Station for Research, Inc.
(iv) To compare an *in situ* XAD-2 pumping system with the other methods of collection and extraction.

**Table II.4**
**Summary of definition of samples obtained during the second cruise for organochlorine sampling.**

| Definition of sampling/processing method used | Sample identification number | Description of origin and history of sample |
|---|---|---|
| XAD-2 extracts from large tank | 800741-747 ; 800751-757 | Water from gaslift system, total use 500 psi. 19/1 :11.15-13.10h until tank full. Extraction with 14 XAD columns then started. 1280 litres extracted. Tank refilled 20/1 :04.00h. Each participant two columns (Table 2). |
| XAD-2 extracts in situ sampling | 800722 ; 773 | 19/1 :15.50-22.00h 74.5 litres water Texas XAD column sample 773<br>19/1 :22.30-05.15 (20/1) 64 litres Monaco column sample 772. |
| Glass flasks, direct sampling | 800774 | 3 individual casts, 10 litres each. 19/1 :9.45-10.15h, extracted immediately idem 17.45-18.15h, extracted imm. Extracts combined 60 litres extracted. |
| Glass flasks from gaslift system directly | 800775 | 3 10-litre flasks filled from gaslift system 19/1 :22.00-22. 30h, extracted immediately. Repeated 20/1 :04. 30-05.00h. Extracts of 60 litres combined. |
| Glass flasks from Bodman, sampler with time delay | 800777 | Bodman cast 19/1 :16.40h. Water remained in sampler onboard until withdrawn 60 litres into flasks. |
| XAD-2 extract from gaslift system drained into drum no 1 | 800776 | 19/1 :10.40-11.10h. 210 litres from gaslift system into drum no 1. 200 litres extracted with XAD-2 column onboard. |
| XAD-2 extract from Bodman drained into drum no 2 | 800783 | 2 Bodman casts (90 litres each) into drum no 2 19/1 :16.00-16.30. 100 litres from drum extracted with XAD-2 onboard. |
| XAD-2 extracts from glass carboys | 800761, 762, 763 | 8 50-litre glass carboys filled from gaslift system 19/1 :14.00-15.00h. 200 litres withdrawn through three XAD columns in series (761, 762, 763) with time delay : 21.1 : 11.00h. |
| Glass flasks filled from Bodman sampler with time delay | 800777 | Bodman cast (19/1 :16.00-16.30). Water remained onboard in Bodman 60 litres withdrawn into flasks. |
| Glass flasks from glass carboys | 800779 | 8 50 litre glass carboys filled from gaslift system 19/1 :14.00-15.00h. 60 litres withdrawn in flasks and extracted in the period 21/1 :10.15-14.00h. |
| | | 700 litres remaining in the large tank after withdrawal of water for 14 XAD-2 columns to be split between participants. Water remained in tank onboard, was brought ashore on 21/1 :10.00h. |
| Glass flasks | 800769 | Gin drum filled from gaslift system 19/1 :13.20-13.40h. Brought ashore 21.1 :10.00h withdrawn 30 1 23/1 :12.00-15.00h. |
| XAD-2 extract | 800701 | 420 litres water drained from one Bodman cast (90 litres) 15/1 : 11.45h and additionally — when that sampler was lost, from gaslift system, operating at about 5 m depth, into 2 stainless steel drums of 210 litres each.<br>Extraction with XAD-2 started immidiately onboard. 125 litres extracted. Extraction took place between 15/1 :14.45 and 16/1 : 00.25h. |
| XAD-2 extract | 800702 | 50 litres extracted from the same drum as above from 14.45h onwards til 18.05h. |
| XAD-2 extract | 800703 | 50 litres extracted from drum as above. |

(v) To perform the standard Panulirus Station hydrographic cast.

(vi) To make a Niskin bottle cast to 900 m for nutrient, dissolved oxygen, and chlorophyll samples.

For storage of the sea-water samples, eight 50-litre glass carboys were cleaned with chromic acid, concentrated hydrochloric acid, water, acetone and hexane. The last hexane washing was evaporated down to an appropriate volume and analysed by gaz chromatography to check the cleanliness of the carboys.

A 1000-litre steel tank, supplied by the Bermuda Biological Station for Research, Inc., which was steam-cleaned and washed with sea water, was used to store sea water obtained through the gas-lift system, so as to minimize any inhomogeneity that might arise when using individual glass carboys as originally proposed in the operational plan.

The samples for intercalibration between the participants were prepared on board the R.V. GEORGE B. KELEZ from sea water collected in the large tank, from the stainless-steel drum containing water from the Bodman sampler, and from the stainless-steel drum containing sea water from the gas-lift system. The effects of storage of sea water in various containers were minimized by immediate extraction through XAD-2 column on board the ship.

For comparison, hexane extracts of sea water were obtained directly from glass sphere water samplers provided by the German Hydrographic Institute in Hamburg (Fig. II.2)

All of the above mentioned objectives were achieved (Table II.4) except the Nansen cast, which had to be cut short owing to deteriorating weather conditions. In achiev-

**Table II.5**
**Identification of samples obtained by extracting water from large tank with individual XAD-2 columns.**

| Laboratory* | Analysis in home lab | | Analysed at Sermuda | |
|---|---|---|---|---|
| | No | Litres extracted | No | Litres extracted |
| BBS | 800741 | 100 | 800751 | 100 |
| TAMU | 800742 | 120 | 800752 | 60 |
| DMI | 800743 | 120 | 800753 | 60 |
| BIO | 800744 | 120 | 800754 | 65 |
| ILMR | 800745 | 120 | 800755 | 65 |
| NISR | 800746 | 110 | 800756 | 65 |
| IMR | 800747 | 110 | 800757 | 65 |

* *Identification of laboratories in Table II.7*

ing these objectives, 48 discrete samples were obtained which amounted to approximately 4000 litres of water. Sampling took place during 17 hours out of the 20 hours spent on station.

Considering the multiple objectives and the poor weather, the trip was considered successful.

## II.4 Results Obtained During the Bermuda Workshop

### XAD-2 extracts from gaslift system

Water from the gaslift system was pumped through XAD-2 columns on board ship. Each participant analysed one column at Bermuda. Volumes of water extracted (between 60 and 100 litres) and the identification of sample numbers and participants are given in Table II.5. Some typical capillary-column temperature-programmed chromatograms of the XAD-2 extracts and of Aroclor 1221, 1254 and 1260 technical formulations are given in Figure II.21.

All chromatograms have a similar peak pattern. However, the relative peak intensities vary considerably amongst the chromatograms.

The peak patterns suggest that the composition of PCB covers the range of components present in standard formulations with widely different overall chlorine content, such as Aroclor 1221, 1254 and 1260 (Fig. II.21).

At that stage it was felt that no attempts should be made to quantify PCB in the samples in terms of any standard formulation such as Aroclor 1254, or in terms of any mixture of standard formulations. Instead, solutions of the various individual PCB components, which were obtained at the beginning of the exercise, were prepared in appropriate concentrations and injected in order to obtain retention times on the Hewlett Packard system, so as to allow identification and quantification of individual PCB components in the sample extract. Table II.6 shows which individual components have corresponding peaks in the standard formulations Aroclor 1221, 1254 and 1260 and the various analysed sample extracts.

It should be oserved that early eluting components in the sample-extract chromatograms have retention times characteristic for some di- and tri-chlorobiphenyls.

The most appropriate approach to quantification of PCBs in the samples is by estimating the contribution of individual components rather than the contribution in terms of standard formulation equivalents. At that time the concentration of PCB in terms of Aroclor 1254 equivalent was estimated to be 0.05 and 0.14 μg $l^{-1}$ in two samples on the basis of one individual PCB component. We shall come back to this problem later when discussing the evolution of ideas that have emerged during the intersessional work after Bermuda. Keeping in mind the size of the problems encountered, and consequently the tremendous amount of work and the sophisticated equipment needed to solve the problems, it was decided to attack the problem only qualitatively while at Bermuda, and to leave further work to the discretion of those laboratories willing to give priority to this type of essential work.

Serious problems were met in eliminating interfering peaks. This was due to the fact that the XAD-2 columns had been stored dry. Application of the complete procedure to XAD-2 columns that had not been exposed to sea water resulted in a large number of interfering peaks, especially in the early parts of the chromatograms. Most of the peaks could be removed by Florisil treatment. However, Soxhlet extraction of the columns with redistilled acetonitrile for 48 hours and storing them under methanol, greatly improved the quality of the chromatograms, even without Florisil treatment. Under no condition should the resin be allowed to remain dry. However, chromatograms of the complete blank procedure still contained components with the same retention times as some of the individual PCB components. These are identified in Table II.3 with an asterisk.

### In situ sampling using XAD-2 resin

The two samples obtained by *in situ* sampling using XAD-2 columns (prepared in the IAEA International Laboratory of Marine Radioactivity in Monaco and Texas A&M University) were analysed at Bermuda.

Concentrations of PCB in terms of Aroclor 1254 equivalent using one individual component were 0.03 ng.$l^{-1}$ in both samples. Hexachlorobenzene, lindane and p, p'-DDE were below the level of detection (0.001 ng.$l^{-1}$). Later analysis in the Texas A&M laboratory of an *in situ* XAD-2, 400-litre sample (obtained at the station's dock) resulted in 0.3 ng.$l^{-1}$ lindane, 0.01 ng.$l^{-1}$ p, p'-DDE and 0.09 ng.$l^{-1}$ PCB Aroclor 1254 equivalent.

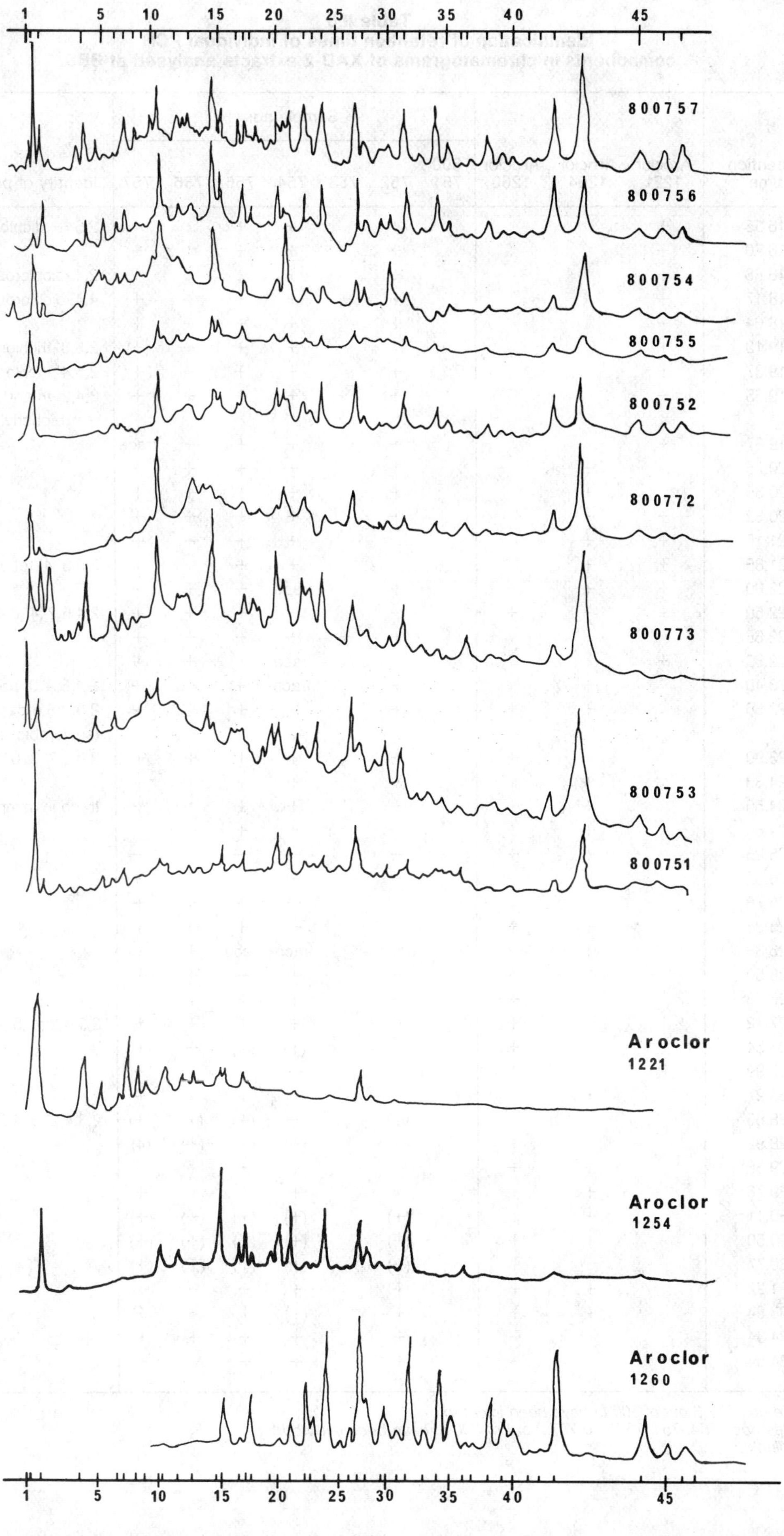


**Figure II.21 Capillary column temperature programmed chromatograms of the XAD-2 extracts of the samples obtained during the first cruise and of Aroclor 1221, 1254 and 1260 technical formulations. Sample identification numbers refer to Table II.5.**

## Table II.6
**Identification of retention times of individual PCB components in chromatograms of XAD-2 extracts analysed at BBS.**

| Peak no | Retention time | Aroclor 1221 | Aroclor 1254 | Aroclor 1260 | Sample nos 800 751 | 752 | 753 | 754 | 755 | 756 | 757 | Identity of peaks |
|---|---|---|---|---|---|---|---|---|---|---|---|---|
| 102* | 16.35 | + | | | | + | | + | + | | + | 2,5 — dichlorobiphenyl |
| 102 | 16.70 | + | | | | + | | + | + | | + | |
| 103 | 16.88 | + | | | | + | | + | + | + | + | 2,4'-dichlorobiphenyl |
| 104* | 18.17 | + | | | | + | + | + | + | + | + | 4,4'-dichlorobiphenyl |
| 105* | 18.64 | + | | | | + | | + | + | + | + | |
| 106* | 19.12 | + | | | | + | | + | + | + | (+) | 2,5,3'-trichlorobiphenyl |
| 107 | 19.37 | + | | | | + | | + | + | + | + | 2,5,4'-trichlorobiphenyl |
| 108* | 19.65 | + | | | | + | | + | + | + | + | 3,4,2'-tri and 2,3,4'-trichloro-biphenyl |
| 109 | 19.87 | + | | | | + | | + | + | + | trace | |
| 110 | 20.25 | | + | | | + | | + | + | + | + | |
| 111* | 20.35 | | + | | | + | | + | + | + | + | |
| 112 | 20.32 | + | + | | | + | | + | + | + | + | |
| 113 | 21.15 | + | + | | | | | + | + | + | + | |
| 114* | 21.85 | + | + | | | + | | + | + | + | + | 2,5,3',4'-tetrachlorobiphenyl |
| 115 | 22.00 | + | + | + | | + | | trace | + | + | + | |
| 116 | 22.50 | + | + | + | | + | | + | + | + | + | 2,4,6,2',4',6'-hexachlorobiphenyl |
| 117 | 22.66 | | + | + | | + | + | + | + | + | + | |
| 118 | 22.83 | + | | | | | | trace | + | + | + | |
| 119 | 23.40 | | + | — | | — | | trace | + | + | + | 2,4,5,2',3'-pentachlcrobiphenyl |
| 120 | 23.56 | | + | + | | + | | + | + | + | + | 2,3,4,5,6-penta 2,3,4,2',5'-pentachlorobiphenyl |
| 121 | 23.90 | | + | + | | + | | + | + | + | + | 2,3,6,2',3',6'-hexachlorobiphenyl |
| 122 | 24.33 | | + | + | | + | | + | + | + | + | |
| 123 | 24.50 | | + | + | | + | | + | + | + | + | trace in samples |
| 124* | 24.80 | | + | + | | — | | + | + | + | + | |
| 125 | 25.25 | | + | + | | — | | — | — | — | — | |
| 126 | 25.52 | | + | + | | — | | — | — | + | + | |
| 127 | 25.78 | | + | + | | + | | — | + | + | + | |
| 128 | 25.98 | | + | + | | + | | — | + | + | + | |
| 129 | 26.38 | | + | + | | trace | | trace | trace | + | (+) | |
| 130* | 26.50 | | | + | | + | | + | + | + | + | |
| 131 | 26.73 | | + | + | | + | | + | + | + | + | |
| 132 | 27.12 | | + | + | | + | | + | + | + | + | 2,3,4,2',4',5',-hexachlorobiphenyl |
| 133 | 27.54 | | + | + | | + | | (+) | (+) | + | + | |
| 134 | 27.99 | | + | + | | + | | — | + | + | + | |
| 135* | 28.27 | | + | + | | + | | + | + | + | + | |
| 136 | 28.66 | | + | + | | (+) | | — | (+) | (+) | (+) | 2,3,4,2',3',4',-hexachlorobiphenyl |
| 137 | 28.92 | | | + | | (+) | | (+) | (+) | (+) | (4) | |
| 138 | 29.38 | | + | + | | + | | + | + | + | + | |
| 139 | 29.76 | | + | + | | + | | + | + | + | + | |
| 140 | 30.11 | | + | + | | (+) | | (+) | (+) | (+) | (+) | |
| 141 | 30.50 | | + | + | | (+) | | (+) | (+) | (+) | (+) | |
| 142 | 30.77 | | | + | | (+) | | (+) | (+) | (+) | (+) | |
| 143 | 31.22 | | + | + | | + | | + | + | + | + | |
| 144* | 33.64 | | + | + | | + | | (+) | + | + | + | |
| 145* | 34.34 | | | + | | + | | + | + | + | + | |
| 146* | 34.84 | | | + | | + | | + | + | + | + | |

*For sample no : 752,3 out of 200 µl have been injected.*
*For samples nos : 754, 755, 756 and 757,2 out of 300 µl have been injected.*
** Also in blank.*

**Table II.7**
**Outline of the methods used at the home laboratories for the quantitation of organochlorine compounds in XAD-2 extracts of seawater.**

| Acronym | Institute | Pre-GLC separation | Clean-up | Packed column analyses | Capillary column analyses |
|---|---|---|---|---|---|
| BIO | Chemical Ocean Div. Atlantic Ocean Lab. Fish & Envir. Canada Bedford Institute of Oceanography, BIO P.O. Box 1006 Dartmouth, N.S. B2Y 4A2, Canada | None | Florisil | | OV 101 25 m temp. prog. 7°C for 2 min. 4°C/min. to 230°C Make-up — Methane Argon 50 ml-min$^{-1}$ |
| TAMU | Department of Chem. and Ocean. Texas A & M Univ. TAMU College Station Texas, USA | None | Florisil | | SE-52 30 m temp. prog. 150°C for 5 min. 4°C/min. to 210°C, isoterm for 5 min. flow rate 4mL $H_2$/min. and 25mL make-up gas Argon/5 % methane |
| IMR | Inst. of Marine Research IMR Direct. of Fisheries P.O. Box 1870 5011 Bergen — Nordnes, Norway | None | $H_2SO_4$ conc. | | SE-54 50 m temp. prog. 100-230°C 8°C/min. flow rate helium 1.5 ml. min$^{-1}$ Make-up nitrogen 50 ml. min$^{-1}$ t° injector 250°C t° detector 250°C |
| DHI | German Hydrogr. Institute DHI Bernhard Noch St. 78 2000 Hamburg Germany | H.P.L.C. | Florisil $A1_2O_3$ | 11 % OV 17 + QFI on gas chrom. Q 80/100 t° oven : 240°C t° detector : 300°C flow rate : 40mL/min. Argon/Methane | |
| NISR | Netherlands Inst. of sea Research NISR P.O. Box 59 Dan Burg Texel The Netherlands | Separation into PCBs and other chlorinated hydrocarbons using $SiO_2$, micro columns | Chromatography on $A1_2O_3$ micro columns | 1.5 % sp 2250, 1.95 % sp 2401 on supelcoport 100-120 mesh, 6 feet long, isotherm 210°C Detector 280°C Ni 63 Injector 225°C Flow and purge 120 mL/min. nitrogen | SE-30 30 m temp. prog. 60-240°C 8°/min. Make-up — nitrogen 60 mil. min$^{-1}$ Carrier He-gas 16 pri. Splitless injection |
| BBS | Bermuda Biolog. Station BBS St. George's West 1-15 Bermuda | None | Florisil | | SE-54 25 m temp. prog. 70-210°C 8°C/min. |
| ILMR | International Lab. of Mar. Radioactivity ILMR Musée Océanographique Monaco-Ville Monaco | None | Florisil | 10 % DC200 on gaschrom. Q 80-100 t° column 200°C t° injector 210°C t° detector 250°C flow rate 40 mL/min. | Fused silica SP2100 25 m temp. prog. 70-210°C, 8°C/min. t° injector 210°C t° detector 250°C splitless mode flow rate nitrogen 1 ml. min$^{-1}$ Make-up nitrogen 40 ml. min$^{-1}$ |

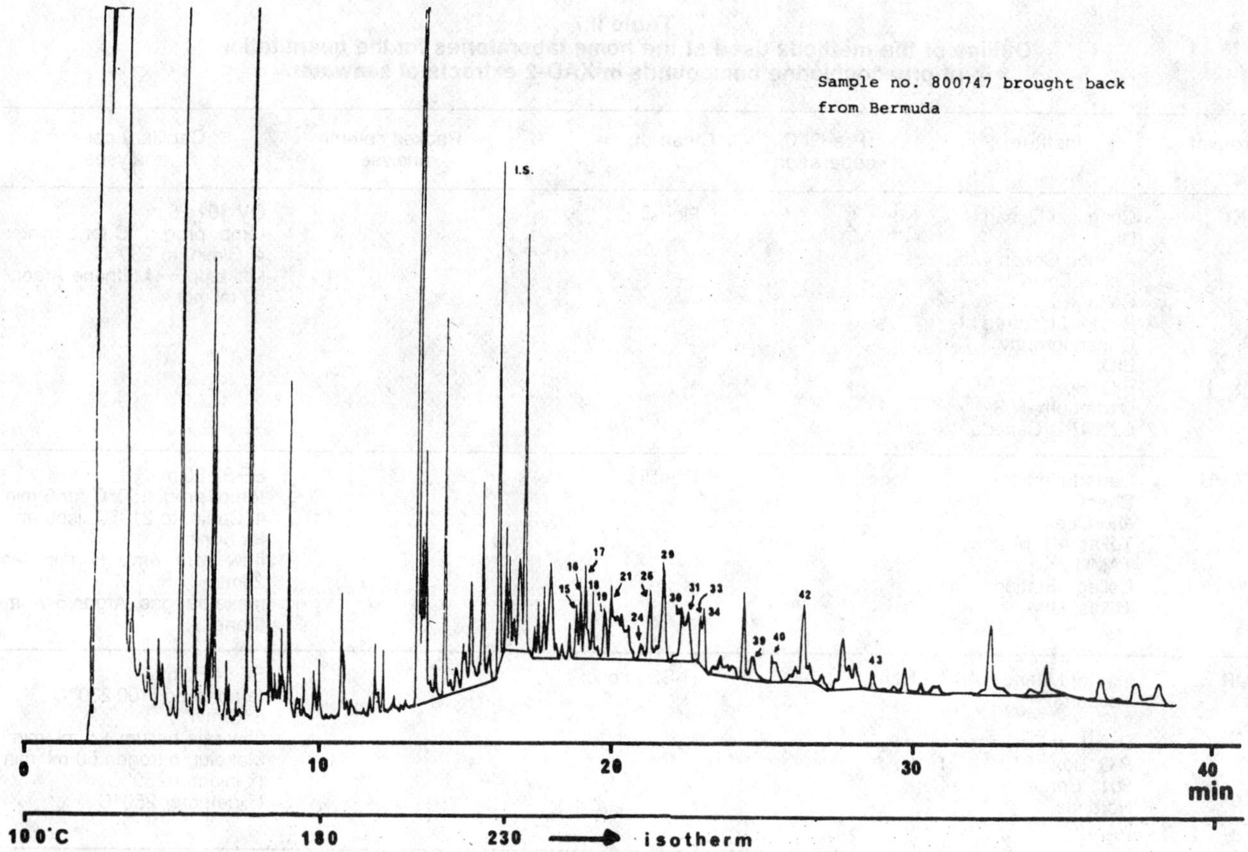


**Figure II.22. Temperature-programmed capillary-column chromatogram of an XAD-2 extract (800747) analysed in one of the home laboratories.**

### (iii) Sampling and extraction in glass-sphere water samplers

It was considered that the procedure involving solvent extraction (with n-hexane) directly in the sampling device (i.e., 10-litre glass flasks) would minimize the risk of an undesired loss or gain of some of the components being measured. It turned out that all chromatograms of extracts that had been obtained along this line had considerably higher peaks of components with retention times characteristic of PCB components. In addition, several other unidentified peaks were present. All these peaks invariably appeared in chromatograms of the blank procedure. One important factor was that a relatively large volume of solvent was used for extraction. Any impurity present in the solvent will be relatively important as an interfering compound due to the extremely low organochlorine levels in the sea-water samples (see Table II.8). All efforts while at Bermuda (repeated distillation of solvents and rinsing of glassware) did not result in dramatic improvements. We have not been able to trace the relative importance of solvent impurity and any possible sorption processes at the surface of the glass flasks, taking place over longer periods.

## Results obtained in the Home Laboratories

Extracts of XAD-2 columns taken from Bermuda (identification as in Table II.5) were analysed in the home laboratories between February and July 1980. An outline of the methods used is given in Table II.7.

In most cases the quality of the chromatograms obtained in the home laboratories (Fig. II.22) is better than those obtained at Bermuda (Fig. II.21). The main reason is the larger volume of water extracted (100 and 65 litres respectively, see Table II.5). In addition, the quality of the blanks had been improved as a result of the experiences obtained at Bermuda, and because of good quality solvents being available at the appropriate time. Results have been obtained by packed-column gas chromatography and in addition, several laboratories reported results on the basis of temperature-programmed capillary-column gas chromatography as at Bermuda. This allowed the quantification of individual components. Until now this has been restricted to only a few components. The quantitative results obtained by the different laboratories and the methods used in quantifying the PCB content of the sample are listed in Table II.7. It should be emphasized that identification of the individual PCB components is based on retention times only for the time being. Further work on GC-MS identification techniques is necessary.

On the assumption that the identification of PCB components is correct, the concentrations of some individual components in the sea water sample are in the order of picograms per litre. In those cases where any particular component has been quantified by more than one laboratory, the agreement is surprisingly good, taking into account the fact that the concentration in far below the concentration of PCBs in biological tissue.

Some laboratories, observing a close similarity in the chromatographic patterns of sample extract and a particular standard formulation, have calculated PCB content in the sample in terms of Aroclor 1254 equivalent. In such an approach, PCBs are quantified by comparing the heights (or areas) of corresponding peaks in chromatograms of sample and the selected standard formulation such as Aroclor 1254.

Usually, it is not possible to select a standard formulation that reproduces the relative intensities of the various peaks in the chromatogram of the sample extract exactly. The quantitative results therefore depend on the peak or peaks selected for comparison. Instead of the selection of one peak, the sum of several prominent peaks has been most commonly used for comparison, to compensate for this problem. The application of this approach on the basis of packed-column chromatograms resulted in concentrations of PCB in terms of Aroclor 1254 equivalent in the low or sub ng. $l^{-1}$ range (Table II.8). Similarly, some participants used temperature-programmed capillary-column chromatograms to obtain Aroclor 1254 equivalent concentration data

**Table II.8**
**Results obtained by analyses of extracts of XAD-2 columns analyzed in the home laboratories.**

| Laboratory | Estimation of PCB content in term of a standard formulation equivalent | | | Quantitation of individual components | |
|---|---|---|---|---|---|
| | Based on standard formulation | Number of peaks selected | PCB (ng $l^{-1}$) | PCB component | ng $l^{-1}$ |
| IMR Bergen | Aroclor 1254 | 15* (cap. col.)<br>5* dominant peaks (cap. col.) | range 0.06-1.57<br>average 0.44<br>range 0.09-0.39<br>average 0.25 | 2,5,3',4'-tetrachlorobiphenyl<br>2,4,5,2',3'-pentachlorobiphenyl<br>2,3,4,2',5'-pentachlorobiphenyl<br>2,4,5,2',4',5'-hexachlorobiphenyl<br>2,3,4,2',4',5'-hexachlorobiphenyl<br>2,3,4,2',3',4'-hexachlorobiphenyl | 0.038<br>0.010<br>0.030<br>0.052<br>0.042<br>0.009 |
| NISR, Texel | Clophen A 50 | sum of 7 (packed col.) | 0.4 | 2,5,3',4'-tetrachlorobiphenyl<br>2,4,5,2',3'-pentachlorobiphenyl<br>2,3,4,5,6-pentachlorobiphenyl<br>2,4,5,2',4',5'-hexachlorobiphenyl<br>2,3,4,2',3',4'-hexachlorobiphenyl | 0.029<br>0.002<br>0.012<br>0.021<br>0.003 |
| IAEA, Monaco | Aroclor 1254 | sum of 6 (packed col.)<br>sum of 9 (cap. col.) | 0.14<br>0.10 | 3,4-dichlorobiphenyl<br>2,5,4'-trichlorobiphenyl<br>2,3,4-trichlorobiphenyl<br>2,3,2',3'-tetrachlorobiphenyl<br>3,4,3',4'-tetrachlorobiphenyl<br>2,4,5,2',3'-pentachlorobiphenyl<br>2,4,5,2',4',5'-hexachlorobiphenyl<br>2,3,4,2',4',5'-hexachlorobiphenyl<br>2,3,4,2',3',4'-hexachlorobiphenyl | 0.009<br>0.002<br>0.001<br>0.008<br>0.011<br>0.001<br>0.017<br>0.003<br>0.010 |
| DHI, Hamburg | Clophen A 60 | sum of 5 (packed col.) | 0.12 1st elution<br>0.08 2nd elution<br>0.20 | (Glasswool plug on top of the XAD-2 column was not removed) | |
| BIO, Dartmouth | Aroclor 1254 | sum of 12 (cap. col.) | 0.23 | | |
| TAMU, Texas | Aroclor 1254 | 5* (cap. col.) | range 0.20-0.35<br>average 0.24 | | |
| BBS, Bermuda | Aroclor 1260 | sum of 13 (cap. col.) | 0.5 | | |

** each of the 15 (5) peaks has been used for quantitation of total PCB, X = 0.44 (0.25).*

based on comparison of peaks of individual components. Concentrations calculated for 15 single peaks taking one peak at the time, ranged from 0.06 to 1.57 ng. $l^{-1}$ Aroclor 1254 equivalent (average 0.44). Using the five most dominant peaks, the result ranged from 0.09 to 0.39 ng. $l^{-1}$ (average 0.25 ng. $l^{-1}$ Aroclor 1254 equivalent).

These large ranges reflect the difference in PCB composition in the sea-water sample and Aroclor 1254. Therefore, all these results have to be considered as arbitrary. They do not reliably reflect the PCB composition of sea water. The appropriate approach to the estimations of PCB in environmental samples involves identification and quantification of individual PCB components, rather than estimating PCB in terms of a standard formulation, even if data for only a limited number of components can be obtained (Duinker, Hillebrand, Palmork & Wilhelmsen, 1980a and b).

## II.5 Discussion and conclusions

In this work we have only focused on poly chlorinated biphenyls. They pose the largest analytical problems and other compounds looked for could hardly be traced and identified without the use of GC MS, which was not available at Bermuda.

Literature on PCBs in sea water, being very limited indeed, has reported concentrations until now almost exclusively in terms of Aroclor 1254 (or the like) equivalent. The present report is one of the very first — and to our knowledge the first one by an international body — openly rejecting this method. At the same time, it demonstrates that not only is the method that makes measurements based on individual PCB components, the most scientifically meaningful method, but that it is feasible even for the extremely low levels found in open-ocean sea-water samples.

The success of the intercalibration exercise is attributed to the following :

(i) the presence of the most up-to-date GC instrumentation allowing the application of temperature-programmed glass-capillary gas chromatography, which has been shown through this work to be essential. The presence of representatives of the various companies ensured « state-of-the-art » performance. It was felt that the availability of a device that records retention times while obtaining chromatograms greatly assisted in the interpretation of results. The use of this type of equipment is highly recommended.

(ii) the excellent laboratory facilities and the staff at Bermuda Biological Station for Research, Inc.

(iii) the extensive work performed in laboratories that had volunteered to undertake such work in the period between January and September 1980. The facilities offered for meetings by the IAEA International Laboratory of Marine Radioactivity in Monaco and the Netherlands Institute for Sea Research in Texel in this period also greatly assisted in the rapid evolution of ideas. At Bermuda it was decided to use, as far as possible, temperature-programed-capillary gas chromatography to quantify PCBs in terms of individual components. At that stage, not all the participating laboratories were adequately equipped for this purpose. However, the experience gained at Bermuda, has accelerated in the mean time the acquisition of appropriate equipment by several laboratories.

Analyses of the XAD-2 extracts in the home laboratories have been performed using capillary-column GC by practically all participants, in addition to packed-column analyses by three participants (Tables II.7 and II.8) other operational factors were the easy access to the open-ocean sampling site from Bermuda, the availability of the NOAA research vessel GEORGE B. KELEZ for the sampling activities, a wide range of different sampling devices and the efforts of an international group of chemical oceanographers with experience in organochlorine analyses.

In spite of the operation being considered successful on the whole, the comparison of the various sampling techniques has not been totally realized. This is attributed to several causes.

(a) As pointed out previously, as a consequence of the low levels of PCBs found in the water samples, a fundamental question arose concerning the validity of the way in which PCBs were quantified using some particular standard formulation (i.e., Aroclor 1254). As a result, a redirection in the major thrust of the operation was made, which shifted the emphasis to analytical considerations.

(b) The weather conditions presented severe difficulties for the sampling phase of the experiment. It was not possible to work up one set of samples prior to the next sampling day. In retrospect, a longer period for shipboard activity would have been desirable.

## II.6 Future work

The experience and results obtained from the analyses of the XAD-2 extracts in the home laboratories after the Bermuda experiment allows some final conclusions to be drawn as to feasibility of making reliable estimations of the PCB content of open-ocean waters. Despite the active cooperation of several specialists in the field, some serious problems were encountered, some of which were solved. However, several problems still need considerable research in the near future, in addition to the work already going on in several countries.

(i) The changes in the sample composition that are most difficult to detect and to eliminate probably occur during sampling and other shipboard activities. Further work is necessary to design appropriate techniques for obtaining and handling sea-water samples at the extremely low concentrations of PCB and other, organochlorine compounds found in open-ocean waters. This work will cover sampling devices, routine handling of equipment, clean labs, elimination of sources of contaminants during normal shipboard activities etc...

(ii) It is important to determine the amounts of any particular compound being in solution and in particulate form. It is still not clear to what extent the various techniques used during the exercise are the appropriate means for distinguishing between these forms.

A considerable amount of work in various parts of the world's ocean is necessary to unravel these problems. The solution to these problems is essential before any type of monitoring of PCBs in the ocean, either regionally or globally, can be initiated. The requisite research should be encouraged by the appropriate organizations.

(iii) Further work is necessary to identify the individual PCB components present in sea water from different parts of the ocean, by the application of GC-MS techniques to extracts of large volumes of sea water. Work of this kind in the Member states should be strongly encouraged by IOC.

(iv) The results of the exercise at Bermuda dit not allow a proper comparison of the efficiencies of XAD-2 and other extraction techniques. Follow-up experiments at some laboratories (e.g., the Bermuda Biological Station for Research the Institute of Marine Research, in Bergen, and the Netherland Institute for Sea Research, in Texel) indicate that resin collection and liquid-liquid extraction techniques are both feasible sampling techniques for the analysis of organochlorines in open-ocean waters, resulting in reliable data for the open ocean if appropriate measures are taken to eliminate interfering peaks. This can be checked by running blanks. It is difficult to reach any conclusions, based on the present results, on the comparability of closed and open sampling devices. Those samples collected for solvent extraction were exposed to the atmosphere. However, analysis of the atmospheric content of PCB in the laboratories at Bermuda during the exercise (performed by the Texas A&M University laboratory) shows that the atmosphere cannot have been a significant source of the interfering peaks.

It is felt that a Pilot Project for monitoring organochlorines in open-ocean waters is feasible. It should consist of several phases. It is recommended that a two-year programme be initiated. It should start with the collection of samples by the different techniques in open ocean waters. The Bermuda Biological Station has indicated its interest in this pilot phase. It is highly recommend that the Bermuda Biological Station for Research, Inc. be asked to host the activities. This would allow the use of a locally available smaller ship for a longer period of time in periods of more desirable weather conditions (e.g., April-December).

In the meantime, the Institute of Marine Research (Bergen) has supplied a continuous water extractor (as described by Ahnoff & Josefsson, 1974) to some participants, in order to accelerate research on the comparison of solvent and XAD-2 resin extraction techniques. These attempts should form the basis for an evaluation of the most appropriate methods for sampling and analysis of background levels of organochlorines in open-ocean waters as well as in coastal waters.

The Institute mentioned above also have volunteered to continue the research on extraction in conjunction with the Bermuda Biological Station for Research, Inc. The results obtained after an initial six-month period at Bermuda should be evaluated, possibly including some additional practical work during a two- or three-week workshop of the participants. The results should lead to a formulation of the technique to be applied in the next phase (i.e., an 18-month period of sampling and analysis for organochlorines in waters off Bermuda). The results obtained over this period should be evaluated in close connection with the data on atmospheric deposition, rainfall, etc., that are currently obtained in the acid rainfall programme at Bermuda Biological Station for Research, Inc., and in other programme.

Based on the experience gained, the following recommendations for future activities can be formulated :

(i) Ample time must be allowed for the planning of the exercise, so that all equipment can be transported and participants notified as to the specifics of the programme.

(ii) Participants in such a programme should arrive well before the first sampling day to ensure familiarity with techniques and the host laboratory. Blanks should be run and familiarity with the available instrumentation should be gained.

(iii) The participation of highly trained personnel from the instrument companies whose equipment is being used is essential ; the companies provide « state-of-the-art » analytical equipment, and their personnel ensure maximum performance of the instruments.

(iv) There should be a feedback mechanism between analytical work-up and ship activities, which enables sampling to occur after preliminary analyses. This would allow for modification of experimental design based on hard data.

(v) Co-ordination of the work towards the scientific goals and the normal shipboard activities should be done by the Chief Scientist during the cruise as well as on « lay » days in port. General shipboard activities such as painting and lubricating *must* be controlled in order to prevent contamination of the samples.

(vi) It is essential to have all laboratory equipment on site, with workshop and vessel facilities being able to respond to short-term requirements dictated by laboratory results. The laboratory staff should have experience in open-ocean marine organic chemistry, as was the case in this exercise.

(vii) It is essential that capillary-column temperature-programmed gas chromatography be used so as to obtain maximum separation of peaks in the chromatograms, allowing for the precise and accurate determination of individual PCB components, and other well defined organochlorine compounds.

## Acknowledgments

The results described here could not have been obtained without the utmost care and efforts of the staff (Mr. Harry Barnes and many others) of the Bermuda Biological Station for Research, Inc. ; without the provision by the U.S. National Oceanographic and Atmospheric Administration, of the R.V. George B. Kelez ; without the support, by the U.S. National Science Foundation of the Texas A&M University laboratory ; without the analysis, by Dr. C.S. Giam of the atmospheric samples ; without Analabs Inc., for kindly supplying the chemical standards, without Hewlett-Packard, Varian and Perkin-Elmer, for supplying equipment and skilled personnel ; and without the efforts of the participants we especially recognize and appreciate the work of the officers and crew of the R.V. GEORGE B. KELEZ and of Drs Donald Atwood and George Harvey in the role of Chief Scientist on board. This manuscript was typed by Jessie Pereira and Wenche Meidell, Institute of Marine Research, Bergen.

## References

Ahnoff, M., and Josefsson B., (1974). Simple apparatus for on-site continuous liquid-liquid extraction of organic compounds from natural waters. Anal. Chem. *46*, 658.

Bodman, R.H., Slabaugh, L.V. and Bowen, V.T. (1961). A multipurpose large volume sea water sampler. J. Marine Res. *19*, 141-148.

Duinker, J.C., and Hillebrand M.T.J., (1979). Behaviour of PCB, pentachlorobenzene, hexachlorobenzene, α-HCH, β-HCH, dieldrin, endrin and p.p'-DDD in the Rhine-Meuse estuary and the adjacent coastal area. Neth. J. Sea Res. *13*, 256-281.

Duinker, J.C., Hillebrand, M.T.J. Palmork, K.H. and Wilhelmsen, (1980a). An evaluation of existing methods for quantitation of polychlorinated biphenyls in environmental samples and suggestions for an improved method based on measurement of individual components. Bull. Environm. Contam. Toxicol. *25* : 956-964.

Duinker, J.C., Hillebrand, M.T.J. Palmork, K.H. and Wilhelmsen, S. (1980b). A discussion on the value of the estimation of polychlorinated biphenyls (PCB's) in environmental samples from packed column gas-liquid chromatographic data. Coun. Meet., Int. Coun. Explor. Sea (E : 35) 1-9.

Duinker, J.C. and Hillebrand M.Th.J. (1978). « Minimizing blank values in chlorinated hydrocarbon analyses. » *J. Chrom. 150* : 195-199.

Harvey, G.R. and Steinhauer W.G. and Teal J.M. (1973). Polychlorobiphenyls in North Atlantic Ocean water. Science, *180*, 643-644.

Harvey, G.R. and Steinhauer W.G. (1974). Atmospheric transport of polychlorobiphenyls to the North Atlantic. Atmospheric Environment, *8*, 777-782.

The Pesticide Analytical Manual, vol. I., US Department of Health, Education and Welfare, Food and Drug Administration, Rockville, Md.

IOC (1977). Report of the First Session of the Intergovernmental Oceanographic Commission's Global Investigation of Pollution in the Marine Environment — Group of Experts on Methods, Standards and Intercalibration (IOC/GGE (MSI)-I/3). Paris, 5-9 Septembre 1977. Unesco, 75700 Paris.

IOC (1978). Report of the Second Session of the Intergovernmental Oceanographic Commission's Global Investigation of Pollution in the Marine Environment — Group of Experts on Methods, Standards and Intercalibration (IOC/GGE (MSI)-II/3) (Bergen, Norway, 1-4 May 1978). Unesco, 75700 Paris.

IOC (1980). Report of the Workshop on the Intercalibration of Sampling Procedures of the IOC/WMO/UNEP Pilot Project on Monitoring Background levels of Selected Pollutants in Open-Ocean Waters, Bermuda, 11-26 January 1980. IOC Work-shop Report No. 25. IOC, Unesco, 75700 Paris.

Tokar, J.M., and Harvey, G.R. (1980). A gaslift system for large volume water sampling (to be published).

## List of participants

J. DUINKER
Netherlands Institute of Sea Research
P.O. Box 59
Den Burg, Texel
THE NETHERLANDS

H. GAUL
German Hydrographic Institute
Bernhard-Nocht-Strasse 78
2000 Hamburg
FEDERAL REPUBLIC OF GERMANY

T.J. HILLEBRAND
Netherlands Institute for Sea Research
P.O. Box 59
Den Burg, Texel
THE NETHERLANDS

T. KNAP
Deputy Director
Bermuda Biological Station for Research
St. George's West 1-15
BERMUDA

J. LEONARD
Bedford Institute of Oceanography
P.O. Box 1006
Dartmouth, Nova Scotia
CANADA B2Y 442

K.H. PALMORK
Institute of Marine Research
P.O. Box 1870
N-5011 Bergen-Nordnes
NORWAY

K. SULLIVAN
Chemistry Department
Texas A&M University
College Station, TX 77843
USA

J.P. VILLENEUVE
International Laboratory of Marine Radioactivity
Musée Océanographique
Monaco Ville
PRINCIPAUTÉ DE MONACO

S. WILHELMSEN
Institute of Marine Research
Nordnesparken 2
N-5011 Bergen-Nordnes
NORWAY

# Chapter III
# Intercalibration of organochlorine standard solutions

J.P. VILLENEUVE
International Laboratory of Marine Radioactivity
International Atomic Energy Agency
Musée Océanographique
Monaco

## Abstract

This chapter summarizes the results of an intercalibration exercise of organochlorines in seawater, including the assessment the accuracy and precision of analyses. Two intercalibration mixtures were prepared and distributed to determine analytical variation due to gas chromatographic quantification and seawater extraction procedures. The results demonstrated the latter to be the greatest source of analytical variability. This part of the IOC/WMO/UNEP Pilot Project was executed by the International Laboratory of Marine Radioactivity of the International Atomic Energy Agency (IAEA)

## Introduction

As a part of the preliminary phase of monitoring organochlorines in open ocean waters the following intercalibration exercises were organized :

1. Intercalibration of high level standard solution working standards of a mixture of selected organochlorines. Aliquots of this solution were to be diluted by participating laboratories and used for gas chromatographic quantification of the designated organochlorines to compare with their own working standards.
2. Intercalibration of a spiked seawater sample to check the sampling and extraction procedures used by the participating laboratories. Aliquots of a low level standard mixture were distributed with instructions for adding to a sample of local seawater. Each laboratory then analyzed the spiked seawater by their routine procedures.

## III. 1 Results

### High level standard solution (working standards)

The second session of the GIPME Group of Experts on Methods, Standard and intercalibration (Bergen, 1-4 May 1978) decided that the intercalibration solutions should contain the following compounds :

pp'DDE : 1,1-dichloro-2,2-bis (p-chlorophenyl) ethylene
pp'DDD : 1,1-dichloro-2,2-bis (p-chlorophenyl) ethane
pp'DDT : 1,1,1, trichloro-2,2-bis (p-chlorophenyl) ethane
Lindane γ hexachlorocyclohexane
HCB : hexachlorobenzene
Aroclor 1254 : Commercial mixture of polychlorinated byphenyls (PCB). Aroclor 1254 was chosen as the PCB standard mixture because in ocean surface waters examined to that date, the PCB distribution was most similar to this commercial mixture. The other compounds were selected on the basis of being common contaminants in ocean waters.

Pure compounds were dissolved in undecane to make a working solution. 2.0 ml of this standard mixture were sealed in glass vials and sent to 44 laboratories with instructions for each participant to dilute the vial contents up to 100 ml with their own solvent used for routine analysis. Concentrations in the standard solution were designed to fall in the range of concentrations commonly used by the laboratories for the preparation of their own calibration standards as shown in Table III. 1

**Table III. 1**

**Concentration of organiochlorines in standard solutions**

| Compound | Concentration used for standard solutions (ng ml$^{-1}$) | Concentration in the intercalibration sample after dilution (ng ml$^{-1}$) |
|---|---|---|
| Lindane | 5 — 100 | 20 |
| HCB | 1 — 100 | 10 |
| pp'DDD | 10 — 250 | 200 |
| pp'DDE | 20 — 200 | 100 |
| pp'DDT | 30 — 300 | 300 |
| Aroclor 1254 | 50 — 500 | 500 |

Of the 44 laboratories originally contacted, 7 declined to participate and 5 others replied that they were interested in the programme but had some problems with their instrumentation. Thus, their participation in the programme was cancelled.

The remainding 13 laboratories returned their results. The names and addresses of these laboratories are attached. A summary of the procedures employed in the participating laboratories is given in Table III. 2. Results reported in Table III. 3 are listed by laboratory code number only to permit each laboratory to compare its analyses with the overall averages and with the true values.

To examine the comparability of the data, both Chauvenet's and Dixon's criteria were applied (Table III. 4).

For the high level standard solution, only one outlier for Aroclor 1254 was rejected according to Chauvenet's test (39.5 μg ml$^{-1}$). Without that value all the results obtained yield $\sigma/\bar{x} = < 20$ %, reflecting the good quality of the standards used by participating laboratories.

### Intercalibration of spiked seawater sample

In order to test the variation due to sampling and analytical procedures, one vial containing the selected organochlorine compounds in methanol solution was sent to each laboratory. The dilution procedure to be used was specified as follows : 12 litres of local seawater were collected by each participating laboratory and concentrated by distillation to 10 litres to remove any organic compounds that could interfere with the GC analysis. The contents of the vial were to be added to the 10 litres of « concentrated » sea water with the resulting standard sample to be analysed by the routine techniques used in each laboratory. Vials were sent to 21 laboratories. 11 laboratories have submitted their results. (Names and addresses are attached.) The analytical results are presented in Tables III. 5 and III. 6 along with the expected values for each compound.

## Table III.2
## Description of analytical methods used by participating laboratories

| Lab. Code No. | Extraction Method | Pre-GLC separation | Clean-up | G.C. conditions: Packed column | G.C. conditions: Capillary column |
|---|---|---|---|---|---|
| 2 | | None | Florisil | — | OV 101 25 m. T° progr. 70 °C for 2 min. 4 °C/min. to 220 °C. Make-up Argon-methane 50 ml $min^{-1}$ |
| 8 | 12 l filtered seawater reduced to 10 l by distillation 1 lit. extracted by 3 X 50 ml hexane 3 aliquots used. A 4 th one extracted by continuous liquid-liquid extraction. | None | Alumina | 1.5 % OV 17 + 1.95 % QF1 Gas. chrom. Q 100/120 (lindane & HCB). 3 % SE 30 (other compound). Flow rate : Argon-methane 50 ml/min. T° injector : 250 °C T° oven 220 °C T° detector : 250 °C | — |
| 12 | Extracted twice with 400 ml hexane for 15 min. | — | — | 1.5 m. 5 % SE 30 on chromaton NAHW HMDS. 0.125-0.160 mm. T° injector : 200 °C. T° oven : 180 °C. T° detector : 200 °C. Flow rate nitrogen 25-27 ml/min. through detector : 130 ml $min^{-1}$ | — |
| 13 | « Outline of the Method to be used for the determination of chlorinated hydrocarbons in seawater. | — | — | — | SE 54 50 m. T° progr. 100 °C to 230 °C. 8 °C/min. Flow rate helium 1.5 ml/min. Make-up nitrogen 50 ml/min. T ° injector 250 °C. T° detector : 250 °C. |
| 14 | 10 l seawater extracted by 100 ml in separatory funnel for 15 min. + extraction with 50 ml hexane. | None | $H_2SO_4$ | 3 mm X 1.6. 5 % QFI Gas chrom. Q 80/100. 1.95 + QFI + 1.5 % OV 17 Chromosorb W 80/100. Flow rate 60 ml/min. T° injector : 210 °C T° oven : 180 °C T° detector : 210 °C. | — |
| 15 | 10 l Atlantic water extracted with hexane by stirring 2 X 15 min. | None | $H_2SO_4$ | 3 mm X 2 m. l) 5 % QFI Gas chrom. Q 100/120. 2) 5 % X E60 chromosorb W (AW, DMCS) 100/120. 1) T° injector : 210 °C. T° detector : 210 °C. T° oven : 185 °C. 2) T ° injector : 225 °C. T° oven : 195 °C. T° detector : 210 °C. Flow rate nitrogen 60 ml $min^{-1}$. | — |
| 16 | 10 l of pre-extracted water spiked-extracted by 300 ml hexane by shaking 1/2 h. | HPLC on silica column | Alumina | 2 mm X 3.6 m. 11 % OV 17 + QFI on Gas chrom. Q 80/100. Flow rate Argon-methane 40 ml $min^{-1}$. T° oven : 240°C. T° detector : 300°C. 0.9 % OV 61 + 2.25 % QFI + 0.9 % XE60 on chromosorb 750 80/100. T° oven : 230°C. | — |
| 21 | | | | 1 % Dextril 300. Chromosorb G. 80/100. T° injector : 200°C. T° detector : 215°C Carrier gas : nitrogen. T° column : 200°C. | — |
| 22 | | | *PCB*: Alumina silica gel. *Others*: Florisil 15 % ether in hexane. | *PCB*: 2 % OV 1 chromosorb W 80/100. 3 mm X 2 m. T° column : 167°C. T° detector : 210°C. *Others*: 1.4 % OV 17 + 2.2 % DC-QFI Gas chrom. Q 100/120. 2 % DEGS + 0.5 % $H_3PO_4$. Chromosorb W (AW, DMCS) 60/80. | — |

**Table III.2 (continued**

| Lab. Code No. | Extraction Method | Pre-GLC separation | Clean-up | G.C. conditions Packed column | Capillary column |
|---|---|---|---|---|---|
| 23 | | | Silica gel | 1 % OV 17 Chromosorb W (AW, DMCS) 80/100. | — |
| 24 | | | | 2.2 m — 2 % OV 1 Flow rate nitrogen 60 ml $min^{-1}$. T° oven : 200°C. T° detector : 270°C. | — |
| 26 | « Outline of the method to be used for the determination of chlorinated hydrocarbons in seawater ». | | | — | OV 101 50 m. T° prog. 70°C to 210°C. 8°C min. T° injector : 210°C. T° detector : 250°C. Flow rate nitrogen 1 ml $min^{-1}$. Make-up nitrogen 40 ml $min^{-1}$ |

Examination of the results of the analyses of pp'DDT, pp'DDD, pp'DDE and Lindane yield a variability somewhat greater than for the analyses of the high level standard solution. This situation is true also for Aroclor 1254. If we consider the laboratories which reported values substantially different from the true value, it is observed that 4 are using packed columns as opposed to capillary columns.

It might be recalled that one of the recommendations of the « Workshop on the Intercalibration of Sampling Procedures » held in Bermuda 11 — 26 January 1980 was the necessity to quantify PCB's, not against an industrial standard, but by the individual components, using temperature — programmed capillary column — Gas chromatography employing an election capture detector (ECD).

## III. 2 Conclusions

As stated at the outset of the programme, the aims of these intercalibration exercises were to ensure that participating laboratories are able to detect the selected organochlorines at levels expected in ocean waters with analytical accuracy within 15 % of the true values. Results showed that analytical accuracy was greatest for quantifications of pp'DDD where more than 50 % of the participating laboratories were within the 15 % error margin. For all other analysed compounds, only about one third of the results from the various laboratories meet this goal. Greater variation was seen in the seawater analysis than in the quantification of the standard mixture.

The results obtained during the intercalibration exercises described herein are an improvement over those of previous efforts organized by the Monaco Laboratory in 1974-1975. The previous attempt used homogeneous aliquots of a sample of Amberlite XAD 2 resin charged with organics extracted from Mediterranean seawater. These results are summarized in Table III.7 for comparison.

As one result of the intercalibration of standard solutions, it was hoped to compare the results obtained by the packed column and capillary column used in the quantification of organochlorines by gas chromatography. Due to the low number of laboratories submitting results in the intercalibration exercise it was not possible to make statistical comparisons of the two methods. This problem was addressed at the Bermuda exercise resulting in the recommendation that PCBs be quantified by comparison with individual PCB compounds. This is best accomplished by gas chromatographs equipped with glass capillary temperature programmed systems.

## Table III.3

**Results of the intercalibration of high level standard solution (results in $\mu g\ ml^{-1}$ in the vial received by the participants).**

| Laboratory code number | Aroclor 1254 | pp'DDT | pp'DDT | pp'DDE | HCB | Lindane |
|---|---|---|---|---|---|---|
| 2 | 39.5 | 17.0 | 13.6 | 6.8 | 0.57 | 0.91 |
| 8 | 27.25 | 14.5 | 10.75 | 6.25 | 0.47 | 1.0 |
| 9 | 28.0 | 16.0 | 10.5 | 6.5 | 0.45 | 1.0 |
| 12 | N.R. | 14.5 | 11.0 | 5.5 | N.R. | 1.05 |
| 13 | 30.268 | 16.719 | 12.232 | 5.668 | 0.581 | 0.876 |
| 14 | 23.0 | 12.5 | 7.5 | 5.0 | 0.35 | 0.9 |
| 15 | 22.5 | 12.5 | 7.25 | 4.75 | 0.375 | 0.9 |
| 16 | 25.65 | 16.3 | 9.5 | 3.25 | 0.422 | 0.97 |
| 21 | 29.6 | 13.15 | 9.8 | 5.4 | 0.455 | 1.05 |
| 22 | 16.5 | 12.6 | 9.0 | 5.0 | 0.5 | 1.0 |
| 23 | 18.0 | N.R. | N.R. | N.R. | N.R. | N.R. |
| 24 | 23.5 | 15.0 | 10.0 | 4.8 | N.R. | 0.95 |
| 26 | 26.5 | 16.0 | 9.5 | 4.0 | 0.62 | 1.05 |
| Mean | 25.9 | 14.7 | 10.1 | 5.2 | 0.48 | 0.97 |
| Standard deviation | ±6.0 | ±1.7 | ±1.8 | ±1.0 | ±0.09 | ±0.06 |
| (%) | (23 %) | (12 %) | (18 %) | (19 %) | (19 %) | (6 %) |
| Expected value | 25.0 | 15.0 | 10.0 | 5.0 | 0.5 | 1.0 |

**Table III.4**

**High level standard solution.**

| | Aroclor 1254 | pp'DDT | pp'DDD | pp'DDE | HCB | Lindane |
|---|---|---|---|---|---|---|
| No. of laboratories participating | 13 | | | | | |
| No of results (N) | 12 | 12 | 12 | 12 | 10 | 12 |
| Maximum value | 39.5 | 17.0 | 13.6 | 6.8 | 0.581 | 1.05 |
| Minimum value | 16.5 | 12.5 | 7.25 | 3.25 | 0.35 | 0.876 |
| Overall average (X) | 25.9 | 14.7 | 10.1 | 5.2 | 0.48 | 0.97 |
| σ* | ±6.0 | ±1.7 | ±1.8 | ±1.0 | ±0.09 | ±0.06 |
| (%) $(\frac{\sigma}{\bar{X}}\ \frac{T}{\bar{X}} \times 100)$ (%) | (23 %) | (12 %) | (18 %) | (19 %) | (19 %) | (6 %) |
| No of results after Chauvenet's test | 11 | 12 | 12 | 12 | 10 | 12 |
| Range | 16.5 — — 30.268 | 12.5 — — 17.0 | 7.25 — — 13.6 | 3.25 — — 6.8 | 0.35 — — 0.581 | 0.876 — — 1.05 |
| Average | 24.6 | 14.7 | 10.1 | 5.2 | 0.48 | 0.97 |
| σ* | ±4.4 | ±1.7 | ±1.8 | ±1.0 | ±0.09 | ±0.06 |
| (%) | (18 %) | (12 %) | (18 %) | (19 %) | (19 %) | (6 %) |
| No of results after Dixon's test | 12 | 12 | 12 | 12 | 10 | 12 |
| Range | 16.5 — — 39.5 | 12.5 — — 17.0 | 7.25 — — 13.6 | 3.25 — — 6.8 | 0.35 — — 0.581 | 0.876 — — 1.05 |
| Average | 25.9 | 14.7 | 10.1 | 5.2 | 0.48 | 0.97 |
| σ* | ±6.0 | ±1.7 | ±1.8 | ±1.0 | ±0.09 | ±0.06 |
| (%) | (23 %) | (12 %) | (18 %) | (19 %) | (19 %) | (6 %) |

* *standard deviation* $\sigma = \sqrt{\frac{(\chi_i - \bar{\chi})^2}{N-1}}$ *This equation is applicable to Tables 1 — 5*

## Table III.5

**Results of the intercalibration of spiked sea water sample.**
**(résults in ng l-1)**

| Laboratory code number | Aroclor 1254 | pp'DDT | pp'DDD | pp'DDE | HCB | Lindane |
|---|---|---|---|---|---|---|
| 2 | 6.8 | 3.8 | 2.9 | 1.3 | 0.15 | 0.25 |
| 8 | 4.08 | 2.41 | 1.67 | 0.781 | 0.051 | 0.179 |
| 12 | 0.85 | 0.85 | 1.0 | 1.0 | N.R. | 0.9 |
| 13 | 14.0 | 5.81 | 4.3 | 1.2 | 0.2 | 0.33 |
| 14 | 7.0 | 3.4 | 2.1 | 1.3 | 0.1 | 0.4 |
| 15 | 7.5 | 3.6 | 2.2 | 1.5 | 0.1 | 0.3 |
| 16 | 4.8 | 2.53 | 1.33 | 0.57 | 0.08 | 0.22 |
| 22 | 10.6 | 3.53 | 1.79 | 0.93 | 0.32 | 0.32 |
| 23 | 15.8 | 3.12 | 2.1 | 1.0 | 0.22 | 0.19 |
| 24 | 9.4 | 3.70 | 1.83 | 0.97 | N.R. | N.R. |
| 26 | 4.3 | 4.1 | 2.8 | 1.7 | 0.1 | 0.2 |
| Mean | 7.7 | 3.4 | 2.2 | 1.1 | 0.15 | 0.33 |
| Standard deviation | ±4.4 | ±1.2 | ±0.9 | ±0.3 | ±0.09 | ±0.21 |
| (%) | (57 %) | (35 %) | (41 %) | (27 %) | (60 %) | (64 %) |
| Expected value | 5.0 | 3.0 | 2.0 | 1.0 | 0.1 | 0.2 |

**Table III.6**
**Spiked sea water sample.**

| | Aroclor 1254 | pp'DDT | pp'DDD | pp'DDE | HCB | Lindane |
|---|---|---|---|---|---|---|
| No of participating laboratories | | | 12 | | | |
| No of results | 11 | 11 | 11 | 11 | 9 | 10 |
| Maximum value | 15.8 | 5.81 | 4.3 | 1.7 | 0.32 | 0.9 |
| Minimum value | 0.85 | 0.85 | 1.0 | 0.57 | 0.051 | 0.179 |
| Overall average | 7.7 | 3.4 | 2.2 | 1.1 | 0.15 | 0.33 |
| σ | ±4.4 | ±1.2 | ±0.9 | ±0.3 | ±0.09 | ±0.21 |
| (%) | (57%) | (35%) | (41%) | (27%) | (60%) | (64%) |
| No of results after Chauvenet's test | 11 | 9 | 10 | 11 | 8 | 9 |
| Range | 0.85 — —15.8 | 2.41 — —4.1 | 1.0 — —2.9 | 0.57 — —1.7 | 0.051 — —0.22 | 0.179 — —0.33 |
| Average | 7.7 | 3.4 | 2.0 | 1.1 | 0.13 | 0.27 |
| σ | ±4.4 | ±0.6x | ±0.6 | ±0.3 | ±0.06 | ±0.08 |
| (%) | (57%) | (18%) | (30%) | (27%) | (46%) | (30%) |
| No of results after Dixon's test | 11 | 9 | 11 | 11 | 9 | 9 |
| Range | 0.85 — —15.8 | 2.41 — —4.1 | 1.0 — —4.3 | 0.57 — —1.7 | 0.051 — —0.32 | 0.179 — —0.33 |
| Average | 7.7 | 3.4 | 2.2 | 1.1 | 0.15 | 0.27 |
| σ | ±4.4 | ±0.6 | ±0.9 | ±0.3 | ±0.09 | ±0.08 |
| (%) | (57%) | (18%) | (41%) | (27%) | (60%) | (30%) |

**Table III.7**

**Organochlorine compound concentrations in Amberlite XAD-11 resin (AB-M-1), concentrations in ng.ml$^{-1}$ (ppb) dry resin*.**

**(Intercalibration Exercise 1974-75).**

| Compound | Lindane | pp'DDT | pp'DDE | Aroclor 1254 |
|---|---|---|---|---|
| No. of participating laboratories | | 19 | | |
| No. of results reported | 12 | 4 | 4 | 7 |
| Maximum value | 229 | 270 | 6.9 | 1615 |
| Minimum value | 0.9 | 0.5 | 0.3 | 11 |
| Overall average | 26 | 73 | 2.7 | 260 |
| σ | ±64 | ±130 | ±3.1 | ±600 |
| (%) | (250%) | (180%) | (110%) | (230%) |

** J.P. Villeneuve, M. Marchand, D. Elder, S.W. Fowler, J. La Rosa, E.K. Duursma, D. Vas and P. Parsi in Activities of the International Laboratory of Marine Radioactivity — 1976 Report — pp. 99-106.*

## LIST OF PARTICIPATING LABORATORIES

**Part 1 : High level standard solution**

Chemical Oceanography Division
Atlantic Oceanographic Laboratory
Fisheries and Environment Canada
Bedford Institute of Oceanography
P.O. Box 1006
Dartmouth, N.S. B2Y 4A2
CANADA

Marine Analytical Chemistry Laboratory
Atlantic Regional Laboratory
National Research Council of Canada
1411 Oxford Street
Halifax, N.S. B3H 3Z1
CANADA

Department of Chemistry and Oceanography
Texas A & M University
College Station, Texas 77843
U.S.A.

Institute of Biology of the Southern Seas of the Academy of Sciences of the Ukrainian SSR
Nahimov Prospect 2
Sebastopol 335000
USSR

Institute of Marine Research
Directorate of Fisheries
P.O. Box 1870
5011 Bergen — Nordnes
NORWAY

Institut für Meereskunde der ADW der DDR
2530 Rostock — Warnemuende
GERMAN DEMOCRATIC REPUBLIC

Department of Agriculture and Fisheries for Scotland, DAFS
Freshwater Fisheries Laboratory
Faskally, Pitlocry, Perthshire
SCOTLAND

Akademie der Wissenschaften der DDR
Forschungsstelle für Chemische Toxicologie
705 Leipzig — Permoserstrasse 15
GERMAN DEMOCRATIC REPUBLIC

German Hydrographic Institute
Bernhard-Nocht-Strasse 78
2000 Hamburg
FEDERAL REPUBLIC OF GERMANY

Marine Pollution Laboratory
Hydrographic Department
Maritime Safety Agency
3-1 Tsukiji 5 — Chome
Chuo-Ku
Tokyo
JAPAN

Water Quality Management Division
Water Quality Bureau
Environmental Agency
Environment Analytical Center
1-1 Kasumigaseki 3 — Chome
Chiyodo-Ku
Tokyo
JAPAN

Water Quality Management Division
Water Quality Bureau
Environmental Agency
METOCEAN
1-1 Kasumigaseki 3 — Chome
Chiyoda-Ku
Tokyo
JAPAN

International Atomic Energy Agency
International Laboratory of Marine Radioactivity
Musée Océanographique
Monaco-Ville
MONACO

**Part 2 : Spiked sea water sample**

Chemical Oceanography Division
Atlantic Oceanographic Laboratory
Fisheries and Environment Canada

Bedford Institute of Oceanography
P.O. Box 1006
Dartmouth, N.S. B2Y 4A2
CANADA

Marine Analytical Chemistry Laboratory
Atlantic Regional Laboratory
National Research Council of Canada
1411 Oxford Street
Halifax, N.S. B3H 3Z1
CANADA

Institute of Biology of the Southern Seas
of the Academy of Sciences of the
Ukrainian SSR
Nahimov Prospect 2
Sebastopol 335000
USSR

Institute of Marine Research
Directorate of Fischeries
P.O. Box 1870
5011 Bergen — Nordnes
NORWAY

Institut für Meereskunde der ADW der DDR
2530 Rostok — Warnemuende
GERMAN DEMOCRATIC REPUBLIC

Akademie der Wissenschaften der DDR
Forschungsstelle für Chemische Toxicologie
705 Leipzig, Permoserstrasse 15
GERMAN DEMOCRATIC REPUBLIC

German Hydrographic Institute
Bernhard — Nocht Str. 78
2000 Hamburg
Federal Republic of Germany

Marine Pollution Laboratory
Hydrographic Department
Maritime Safety Agency
3-1 Tsukiji 5 — Chome
Chuo-Ku
Tokyo
JAPAN

Water Quality Management Division
Water Quality Bureau
Environmental Agency
Environment Analytical Center
1-1 Kasumigaseki 3 — Chome
Chiyoda-ku
Tokyo
JAPAN

Water Quality Management Division
Water Quality Bureau
Environmental Agency
METOCEAN
1-1 Kasumigaseki 3 — Chome
Chiyoda-Ku
Tokyo
JAPAN

International Atomic Energy Agency
Internationa Laboratory of Marine Radioactivity
Musée Océanographique
Monaco-Ville
MONACO